MANAGING QoS IN MULTIMEDIA NETWORKS AND SERVICES

IFIP - The International Federation for Information Processing

IFIP was founded in 1960 under the auspices of UNESCO, following the First World Computer Congress held in Paris the previous year. An umbrella organization for societies working in information processing, IFIP's aim is two-fold: to support information processing within its member countries and to encourage technology transfer to developing nations. As its mission statement clearly states,

IFIP's mission is to be the leading, truly international, apolitical organization which encourages and assists in the development, exploitation and application of information technology for the benefit of all people.

IFIP is a non-profitmaking organization, run almost solely by 2500 volunteers. It operates through a number of technical committees, which organize events and publications. IFIP's events range from an international congress to local seminars, but the most important are:

- The IFIP World Computer Congress, held every second year;
- open conferences;
- working conferences.

The flagship event is the IFIP World Computer Congress, at which both invited and contributed papers are presented. Contributed papers are rigorously refereed and the rejection rate is high.

As with the Congress, participation in the open conferences is open to all and papers may be invited or submitted. Again, submitted papers are stringently refereed.

The working conferences are structured differently. They are usually run by a working group and attendance is small and by invitation only. Their purpose is to create an atmosphere conducive to innovation and development. Refereeing is less rigorous and papers are subjected to extensive group discussion.

Publications arising from IFIP events vary. The papers presented at the IFIP World Computer Congress and at open conferences are published as conference proceedings, while the results of the working conferences are often published as collections of selected and edited papers.

Any national society whose primary activity is in information may apply to become a full member of IFIP, although full membership is restricted to one society per country. Full members are entitled to vote at the annual General Assembly, National societies preferring a less committed involvement may apply for associate or corresponding membership. Associate members enjoy the same benefits as full members, but without voting rights. Corresponding members are not represented in IFIP bodies. Affiliated membership is open to non-national societies, and individual and honorary membership schemes are also offered.

MANAGING QoS IN MULTIMEDIA NETWORKS AND SERVICES

IEEE / IFIP TC6 — WG6.4 & WG6.6
Third International Conference on Management of Multimedia Networks and Services (MMNS'2000)
September 25–28, 2000, Fortaleza, Ceará, Brazil

Edited by

José Neuman de Souza
Computer Science Department
Federal University of Ceará
Brazil

Raouf Boutaba
Department of Computer Science
University of Waterloo
Canada

SPRINGER SCIENCE+BUSINESS MEDIA, LLC

Library of Congress Cataloging-in-Publication Data

International Conference on Management of Multimedia Networks and Services (3rd:2000: Fortaleza, Ceará, Brazil)

Managing QoS in multimedia networks and services / edited by José Neuman de Souza, Raouf Boutaba.

"IEEE/IFIP TC6 – WG6.4 & WG6.6 Third International Conference on Management of Multimedia Networks and Services (MMNS'2000) September 25-28, 2000, Fortaleza, Ceará, Brazil."

Includes bibliographical references.

ISBN 978-1-4757-5713-2 ISBN 978-0-387-35532-0 (eBook)

DOI 10.1007/978-0-387-35532-0

1. Integrated services digital networks—Congresses. 2. Computer networks—Management—Congresses. 3. Quality control—Congresses. 4. Multimedia systems—Congresses. I. Souza, José Neuman de. II. Boutaba, Raouf. III. Title.

TK5103.75 I58 2000
004.6'068—dc21 00-061079

Originally published by Kluwer Academic Publishers in 2000

Printed on acid-free paper.

TABLE OF CONTENTS

Part Four - Network Programmability

Part Five - Policies and Service Management

Part Six - Advanced techniques for Network Management

Preface

Welcome to the third International Conference on Management of Multimedia Networks and Services (MMNS'2000) in Fortaleza (Brazil)! The first MMNS was held in Montreal (Canada) in july 1997 and the second MMNS was held in Versailles (France) in November 1998. The MMNS conference takes place every year and a half and is aimed to be a truly international event by bringing together researchers and practitioners from all around the world and by organising the conference each time in a different continent/country.

Over the past several years, there has been a considerable amount of research within the fields of multimedia networking and network management. Much of that work has taken place within the context of managing Quality-of-Service in broadband integrated services digital networks such as the ATM, and more recently in IP-based networks, to respond to the requirements of emerging multimedia applications. ATM networks were designed to support multimedia traffic with diverse characteristics and can be used as the transfer mode for both wired and wireless networks. A new set of Internet protocols is being developed to provide better quality of service, which is a prerequisite for supporting multimedia applications. Multimedia applications have a different set of requirements, which impacts the design of the underlying communication network as well as its management. Several QoS management mechanisms intervening at different layers of the communication network are required including QoS-routing, QoS-based transport, QoS negotiation, QoS adaptation, FCAPS management, and mobility management. These QoS-related issues are grouped under the umbrella of QoS management and covered by MMNS'2000 in the context of multimedia networks and services.

One of the objectives of MMNS is to stress the need for integrated control and management traditionally addressed by separate research communities. Such integration along with network programmability are particularly critical in the context of multimedia networking and QoS management, and facilitate the faster deployment of new multimedia services.

Many conferences on networking offer technical sessions on network management and few conferences on network management offer technical sessions on managing multimedia networks. There is no research forum other than MMNS that provides a single-track conference dedicated to control and management in QoS-sensitive multimedia networks, services and applications.

With this objective in mind, we have selected for the 3rd MMNS the best papers among those submitted. The technical program is composed of 20 presentations based on the papers selected during the reviewing process. Ten countries in four continents are represented in the conference program. Every paper was rigorously reviewed by at least 3 international experts. We think that we have great quality technical sessions each focusing on one facet of managing QoS in multimedia networks and services. The technical sessions are ordered as follows: Performance management in broadband networks; QoS in Internet networks; Multimedia networks and services; Network programmability; Policies and service management; Advanced techniques for network management.

The MMNS'2000 technical program is complemented by an invited program made up of a keynote address, two invited talks and a panel which address timely and hot issues related to the conference technical theme as well as broader subjects.

We would like to thank the authors of the technical papers and the invited speakers, without whom this conference would not have been possible. We would also like to express our gratitude to the program committee and reviewers for their efforts in making this a successful conference. Thanks also go to the Federal University of Ceará for hosting this conference, to Insoft (Istituto do software do Ceará) and to the local organisation committee (particularly, Cris Amon Caminha da Rocha). Last but not the least, we thank the conference sponsors IFIP TC6 WG6.4 & WG6.6, and the IEEE Communications Society as well as Kluwer Academic Publishers (particularly, Yana Lambert).

We wish to all the participants a productive conference and a nice stay in Fortaleza, Ceará, Brazil

José Neuman de Souza and Raouf Boutaba

PROGRAM CO-CHAIRS

José Neuman de Souza, Federal University of Ceará, Brazil
Raouf Boutaba, University of Waterloo, Canada

PROGRAM COMMITTEE

Abdelhakim Hafid, Bellcore, NJ, USA
Ahmed Karmouch, U. of Ottawa, Canada
Andrew Campbell, Columbia University, USA
Dominique Gaiti, U. of Troyes, France
Douglas N. Zuckerman, Telcordia Technologies, USA
Ehab Al-Shaer, DePaul University, Chicago, IL, USA
Emil Lupu, Imperial College, UK
Farouk Kamoun, ENSI, Tunisia
Giovanni Pacifici, IBM T.J.Watson Research Center, USA
Guy Pujolle U. of Versailles, France
Jacques Labetoulle, Eurecom, France
James W. Hong, POSTECH, Korea
Jean Pierre Claudé, U. of Versailles, France
Joaquim Celestino Júnior, UECE, Ceará, Brazil
Joberto S. Martins, UNIFACS, Brazil
Jose M. Nogueira, U. Fed. Minas Gerais, Brazil
Leon A. Garcia, U. of Toronto, Canada
Michael A. Stanton, U. Fed. Fluminense, Brazil
Mohamed Bettaz, Philadelphia University, Amman, Jordan
Nicolas Georganas, University of Ottawa, Canada
Ralph Steinmetz, GMD IPSI & Darmstadt U., Germany
Roberto Saracco, CSELT, Italy
Rolf Stadler, CTR/Columbia U., USA
Salah Aidarous, NEC America, USA
Simon Znaty, ENST-Bretagne, France
Wolfgang Zimmer, GMD FIRST, Germany
Yechiam Yemini, Columbia U., USA

LOCAL ORGANIZATION COMMITTEE

Cris Amon Caminha da Rocha (UFC)
Joaquim Celestino Jr. (INSOFT)
Manoel Orley Mendes Carneiro (UFC)
Marcos Clayton (INSOFT)
Renata Luz (INSOFT)

LIST OF REVIEWERS

Abdelhakim Hafid, Bellcore, NJ, USA
Ahmed Karmouch, U. of Ottawa, Canada
Ahmed Mehaoua, Lab. PRISM, U. of Versailles, France
Andrei Dragoi, U. of Waterloo, Canada
Andrew Campbell, Columbia University, USA
Biswaroop Mukherjee, U. of Waterloo, Canada
Carlos de Castro Goulart, U. Fed. de Viçosa, MG, Brazil
David Evans, U. of Waterloo, Canada
Dimitrios Makrakis, U. of Ottawa, Canada
Dominique Gaiti, U. of Troyes, France
Dorgival Olavo Guedes Neto, U. Fed. Minas Gerais, Brazil
Douglas N. Zuckerman, Telcordia Technologies, USA
Ehab Al-Shaer, DePaul University, Chicago, IL, USA
Emil Lupu, Imperial College, UK
Farag M. Sallabi, U. of Ottawa, Canada
Farouk Kamoun, ENSI, Tunisia
Fethi Filali, INRIA-Sophia Antipollis, France
Giovanni Pacifici, IBM T.J.Watson Research Center, USA
Guillaume Mocquard, U. of Waterloo, Canada
Guy Pujolle U. of Versailles, France
Hajer Tounsi, Ecole Nationale des Sciences de l'Informatique, Tunisia
Hella Kaffel Ben Ayed, Faculté des Sciences de Tunis, Tunisia
Iraqi Youssef, U. of Montreal, Canada
Jacques Labetoulle, Eurecom, France
James W. Hong, POSTECH, Korea
Joaquim Celestino Júnior, UECE, Ceará, Brazil
Joberto S. Martins, UNIFACS, Brazil
Jose M. Nogueira, U. Fed. Minas Gerais, Brazil
Jose Neuman de Souza, U. Fed. do Ceara, Brazil
Jose R. Gallardo, U. of Western, Canada
Lakhdissi Mouhsine, U. of Ottawa, Canada
Leon A. Garcia, U. of Toronto, Canada
Magdi Amer, U. of Ottawa, Canada
Martin McSweeney, U. of Waterloo, Canada
Michael A. Stanton, U. Fed. Fluminense, Brazil
Mohamed Ahmed, U. of Ottawa, Canada
Mohamed Bettaz, Philadelphia University, Amman, Jordan
Mohamed-Vall M. Salem, U. of Montreal, Canada
Nélson Fonseca, U. of Campinas, SP, Brazil
Nicolas Georganas, University of Ottawa, Canada
Ralph Steinmetz, GMD IPSI & Darmstadt U., Germany

Raouf Boutaba, University of Waterloo, Canada
Roberto Saracco, CSELT, Italy
Rolf Stadler, CTR/Columbia U., USA
Salah Aidarous, NEC America, USA
Sami Tabbane,Ecole Supérieure des Communications de Tunis, Tunisia
Sidd Shenoy, U. of Waterloo, Canada
Simon Znaty, ENST-Bretagne, France
Wolfgang Zimmer, GMD FIRST, Germany
Yanni Ellen Liu, U. of Waterloo, Canada

PART ONE

Performance Management in Broadband Networks

Chapter 1

PERFORMANCE EVALUATION OF THE HIERARCHICAL HANDOFF SCHEME FOR WIRELESS ATM NETWORKS

[*]Nadjia Kara, [&]Jean Conan, [#]André Girard
**ECE Departement, Ecole polytechnique, Montreal, Canada, kara@inrs-telecom.uquebec.ca*
&ECE Departement, Ecole polytechnique, Montreal, QC, Canada, conan@com.polymtl.ca
#INRS-Telecom, Université du Québec, Québec, QC,Canada. andre@inrs-telecom.uquebec.ca

Key words: Handoff, Wireless ATM.

Abstract: One of the most important issues in wireless networks is handoff. Recently, a hierarchical protocol has been proposed to manage handoff in wireless ATM networks based on the partial rerouting strategy and the virtual connection tree concept. In this paper, a discrete time model is proposed to estimate the connection bandwidth and the connection handoff delay of this scheme. We propose to compare the hierarchical protocol with the Akyol and Cox handoff protocol. Numerical results illustrate that for the future wireless ATM networks the hierarchical model gives lower connection bandwidth and lower connection handoff delay than those of Akyol and Cox model.

1. INTRODUCTION

Two features are dominating in the development of future communication networks : ATM technology with its high capacity and service capabilities and guaranteed Quality of Service (QoS) on the one hand, and support of mobility on the other hand. The combination of these features is wireless ATM. Wireless ATM is expected to play a significant role in the next generation of wireless broadband networks (Raychaudhuri & Wilson, 1994; The ATM Forum, 1998; Bush et al., 1997) and since no specific standards are yet specified for wireless ATM, this topic is an active research area. Many aspects of wireless ATM have been studied and mobility including connection rerouting to support handoff is one of the most important ones and is the focus of our work.

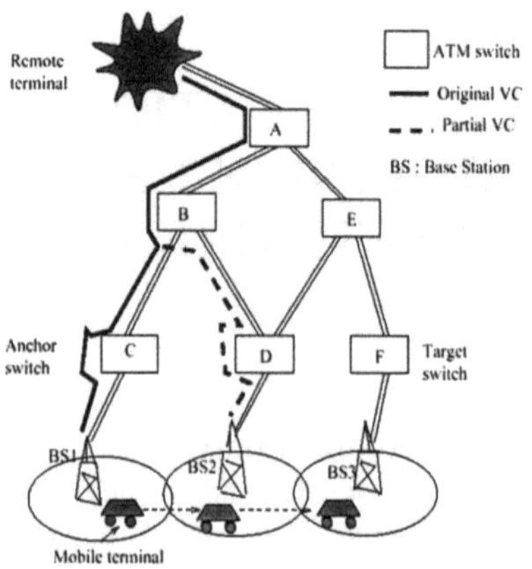

Figure 1. Partial rerouting scheme

Several rerouting protocols have been proposed to simplify handoff management (Acompora & Naghshineh, 1994; Akyol & Cox, 1997; Veerarghavan et al., 1996; Worng & Leung, 1998). One of these is the partial rerouting scheme based on a partial connection re-establishment (Akyol & Cox, 1997). In this case, a switch along the original connection is selected to setup a portion of connection to the target switch to which the new base station is connected (Fig. 1). The switch selected to perform this function is called the crossover switch. This protocol gives an efficient use of bandwidth by setting up a new optimal path. Despite this efficient management of bandwidth, the time to compute this optimal path may exceed the constraint of handoff delay.

Another protocol is the virtual connection tree scheme, where resources are preallocated in the area surrounding the base station to which a mobile terminal is connected (Acampora & Naghshineh, 1994). The tree root is a specialized ATM switch with mobility management capability (Fig. 2). This protocol ensures a fast handoff but the bandwidth use is inefficient.

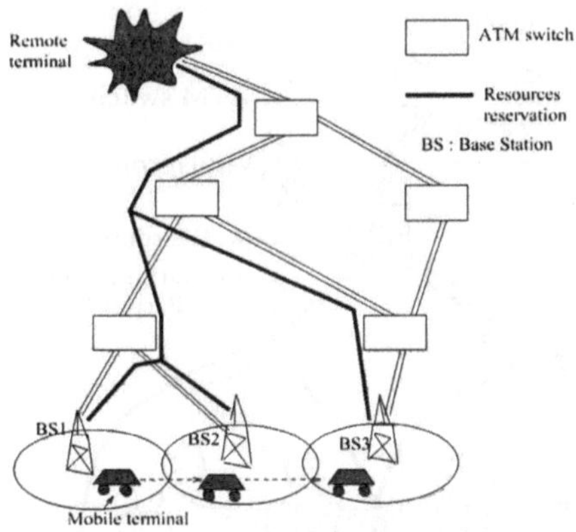

Figure 2. Virtual connection tree scheme

To solve the delay and bandwidth problems, we have proposed in (Kara et al., 1999) a new handoff protocol called hierarchical handoff. In this model, we use trees regrouping base stations and we regroup all of these trees into an arborescence structure connected to the root node. This structure implies two kinds of handoffs : the intra-subnetwork handoff and the inter-subnetworks handoff. The intra-subnetworks handoff is managed by the tree root nodes of the subnetwork. Since in this case the resources are always available, the handoff may occur without delay (for example in figure 3 the tree with root note G for handoff from P_1 to P_2). The inter-subnetworks handoff is managed by using the partial rerouting procedure but only between the root nodes of the subnetworks, for example between C and J for the handoff from P_2 to P_3 (Fig. 3). This handoff scheme allows a distributed mobility management since each subnetwork root node manages the mobility for a set of base stations independently of other root nodes of the subnetwork. The signaling traffic in the network is thus reduced and this reduction is very important for multimedia networks. We expect that this handoff scheme will give a handoff delay and bandwidth smaller than those of the Akyol and Cox handoff scheme.

In this paper, we estimate the bandwidth and handoff delay of the hierarchical model and we compare this performance criteria with those of the Akyol and Cox model. This paper is organized as follows. In section 2, the hierarchical handoff scheme is described and performance analysis of this model is presented where the performance criteria are the bandwidth and handoff delay. In section 3, we describe the performance analysis of the Akyol and Cox handoff scheme by estimating the bandwidth and the handoff delay. In section 4, we present numerical results and the comparison between the two handoff schemes. Conclusion is given in section 5.

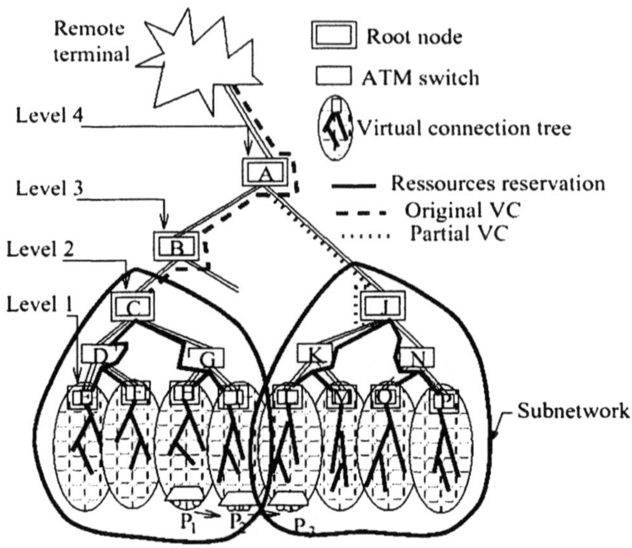

Figure 3. Hierarchical handoff scheme

2. ANALYTICAL MODEL OF THE HIERARCHICAL HANDOFF SCHEME

The hierarchical handoff protocol is a combination of the partial rerouting protocol and the virtual connection tree protocol. When the mobile moves within the subnetwork, intra-subnetwork handoff may occur. In this case, the connection is maintained by the virtual connection tree, and the connection rerouting duration is very short. If the mobile moves outside the subnetwork, then a combined rerouting which consists of a partial rerouting followed by a virtual connection tree rerouting, is performed to maintain the connection. A discrete time model is developed in (Chan & Leung, 1998), for the two-phase handoff scheme, for evaluating the bandwidth of three different path optimization schemes. We propose to adapt and to generalize this model to evaluate the bandwidth and the handoff delay for the hierarchical model. Events that may occur at each time i=1,2,... are : 1) partial rerouting; 2) call termination. The intra-subnetwork handoff is not included because it does not generate signaling traffic. At each time i, we define :

- λ the probability that there is an inter-subnetwork handoff (and thus a partial rerouting);
- μ the probability that the call terminates.

Thus, inter-subnetwork handoff is modeled as a Bernoulli process. The call duration has a geometric distribution with mean $1/\mu$. For each mobile connection, we define :

- L_h the number of links between the root node of the subnetwork to which the mobile belongs and the destination during the call duration.

L_h is a random variable with a general distribution and with mean $\overline{L_h}$. We assume that the number of links of the new connection (rerouted connection) is a random variable with the same mean $\overline{L_h}$.

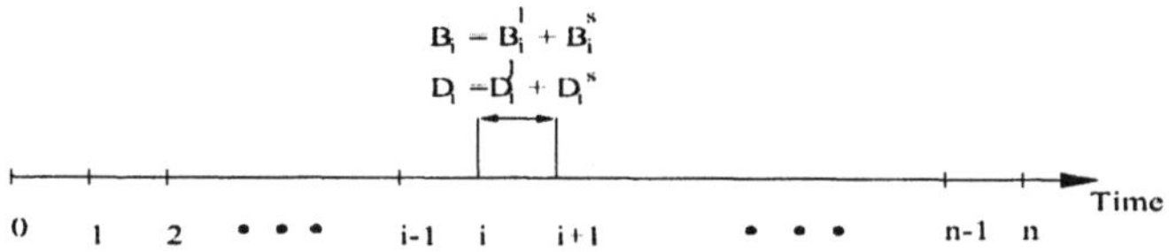

Figure 4. Time diagram

The performance measures are the expected bandwidth per call and the expected handoff delay per call. The expected bandwidth per call is the network bandwidth needed to support a mobile connection, and is the sum of the link bandwidth and the signaling bandwidth due to inter-subnetworks handoff. Likewise, the expected handoff delay per call is the network handoff delay to support a mobile connection, and is the sum of the link delay (delay of user's bandwidth reservation) and the signaling delay due to the inter-subnetworks handoff. Let:

- B be the bandwidth of one call per unit time on a given link;
- D be the delay of the bandwidth reservation of one call per unit time on a given link;
- B_{RP} be the signaling bandwidth for each path rerouting;
- D_{RP} be the signaling delay for each path rerouting.

The parameters B_{RP} and D_{RP} include the bandwidth and the delay respectively for : 1) locating the handoff switch; 2) setting up the new portion of connection; 3) terminating the old portion of connection; and 4) updating the connection server about the status of the existing route.

In the following subsections the problem of determining the expected bandwidth per call and the expected handoff delay per call for the hierarchical handoff scheme is addressed.

By using a discrete time model, we will estimate the expected handoff delay per call denoted $E[D_h]$, and the expected bandwidth per call $E[B_h]$.

The time diagram of figure 4 subdivides the call duration into time intervals. In each time interval $]i,i+1]$ a handoff may occur. In each interval, let:

- D_i be the delay of the bandwidth reservation for one call during time interval $]i,i+1]$ called delay;
- D^l_i be the delay of bandwidth reservation on the links of one call during time interval $]i,i+1]$ called link delay;
- D^l_i be the signaling delay due to handoff of one call during time interval $]i,i+1]$;
- B_i be the bandwidth for one call during time interval $]i,i+1]$ called bandwidth;
- B^l_i be the bandwidth reserved on the links of one call during time interval $]i,i+1]$ called link bandwidth;
- B^S_i be the signaling bandwidth due to handoff of one call during time interval $]i,i+1]$.

Let n be the call termination time. D_i, D^l_i, and D^S_i, are related by the following equation:

$$D_i = D_i^l + D_i^s \tag{1}$$

Likewise B_i, B^l_i, and B^S_i are related by :

$$B_i = B_i^l + B_i^s \tag{2}$$

We define a random variable indicator

$$I_i = \begin{cases} 1 if an\, inter - subnetwork handoff occurs at time i, \\ 0 otherwise \end{cases} \tag{3}$$

Since at each time interval the partial rerouting is modeled as a Bernoulli random variable with probability l, we have $E[I_i]$=l.

2.1 Expected handoff delay per call

The signaling delay of the partial rerouting due to mobility at time i is

$$D_i^s = D_{RP} I_i \tag{4}$$

the expected of D^s is

$$E\left[D_i^s\right] = \lambda D_{RP} \tag{5}$$

and the link delay between time interval $]i,i+1]$ is

$$D_i^l = L_h D I_i \tag{6}$$

The expectation of D^l_i is

$$E\left[D_i^l\right] = \lambda \overline{L_h} D \tag{7}$$

and the expected delay per call is

$$\begin{aligned} &E[D_h] = E[E[D_h / \ N]] = \\ &\sum_{n=1}^{\infty} E\left[\sum_{i=0}^{n-1} D_i \middle| N = n\right] P(N = n) = \sum_{n=1}^{\infty}\sum_{i=0}^{n-1} E\left[\left(D_i^l + D_i^s\right)\right] P(N = n) = \\ &\lambda \frac{1-\mu}{\mu} \overline{L_h} D + \lambda \frac{1-\mu}{\mu} D_{RP} \end{aligned} \tag{8}$$

2.2 Expected bandwidth per call

The signaling bandwidth for partial rerouting due to mobility at time i is

$$B_i^s = B_{RP} I_i \tag{9}$$

the expectation of B^S is

$$E\left[B_i^s\right] = \lambda B_{RP} \tag{10}$$

and the link bandwidth between time interval $]i,i+1]$ is

$$B_i^l = B_{i-1}^l (1 - I_i) + L_h B I_i \tag{11}$$

Taking the expectation of the link bandwidth :

$$E\left[B_i^l\right] = (1-\lambda) E\left[B_{i-1}^l\right] + \lambda \overline{L_h} B \tag{12}$$

the expectation bandwidth per call is :

$$E[B_h] = E\left[E\left[\sum_{i=0}^{n-1} B_i / \ N\right]\right] =$$
$$\sum_{n=1}^{\infty} E\left[\sum_{i=0}^{n-1} B_i \middle| N = n\right] P(N=n) = \sum_{n=1}^{\infty}\sum_{i=0}^{n-1} E\left[B_i^l + B_i^s\right] P(N=n) = \quad (13)$$
$$\frac{1-\mu}{\mu}\overline{L_h} B + \lambda \frac{1-\mu}{\mu} B_{RP}$$

To this bandwidth we add a fixed bandwidth B_{hs} due to the resource reservation in the subnetworks.

$$B_{hs} = L_{hs}\left(B + B_{RP}\right) \quad (14)$$

Where L_{hs} is the number of links in the subnetworks. Since the expected bandwidth per call is

$$E[B_h] = \frac{1-\mu}{\mu}\overline{L_h} B + L_{hs} B + \lambda \frac{1-\mu}{\mu} B_{RP} + L_{hs} B_{RP} \quad (15)$$

the expected handoff delay per call and the expected bandwidth per call of the hierarchical handoff scheme have been estimated. We will now estimate the expected handoff delay per call and the expected bandwidth per call of Akyol and Cox handoff scheme.

3. ADAPTED ANALYTICAL MODEL TO THE AKYOL AND COX HANDOFF SCHEME

The Akyol and Cox handoff protocol is based on the partial rerouting (Akyol & Cox, 1997). At each inter-switch handoff, a partial rerouting is performed. The performance analysis in (Akyol & Cox, 1997) estimates the number of signaling messages and the signaling bandwidth of the protocol. In this section, we adapt the analytical model of the hierarchical handoff scheme to Akyol and Cox handoff scheme to derive the expected handoff delay per call and the expected bandwidth per call and then to compare the hierarchical handoff scheme with the Akyol and Cox handoff scheme.

The rerouting is modeled as the hierarchical rerouting model with the following definitions :

- λ is the probability that there is an inter-switch handoff (and thus a partial rerouting);

– L_S is the number of links between the source and destination.

L_S is a random variable with a general distribution and with mean L_S. We suppose that the number of connection links after the partial rerouting is a random variable with the same mean L_S.

We define a random variable indicator :

$$J_i = \begin{cases} 1 if an inter-switch handoff occurs at time i \\ 0 otherwise \end{cases} \tag{16}$$

At each time interval the partial rerouting is modeled as a Bernoulli random variable with probability λ and $E[J_i]=\lambda$.

In this case, we can compute the expected handoff delay per call and the expected bandwidth per call.

3.1 Expected handoff delay per call

In this case, following the same steps as established for the hierarchical handoff scheme, we arrive at the following results.

The signaling delay of the partial rerouting due to inter-switch handoff at time i is

$$D_i^s = D_{RP} J_i \tag{17}$$

the expected signaling delay due to mobility at time i is

$$E\left[D_i^s\right] = \lambda D_{RP} \tag{18}$$

The link delay between time interval $]i, i+1]$ is

$$D_i^l = L_s D J_i \tag{19}$$

the expected link delay between time interval $]i, i+1]$ is

$$E\left[D_i^l\right] = \lambda \overline{L_S} D \tag{20}$$

The expected handoff delay per call is

$$E\left[D_S\right] = \frac{\lambda(1-\mu)}{\mu}\left(\overline{L_S} D + D_{RP}\right) \tag{21}$$

3.2 Expected bandwidth per call

The signaling bandwidth for partial rerouting due to inter-switch handoff at time i is

$$B_i^s = B_{RP} J_i \tag{22}$$

the expected signaling bandwidth due to mobility at time i is

$$E[B_i^s] = \lambda B_{RP} \tag{23}$$

The link bandwidth between time interval $]i, i+1]$ is

$$B_i^l = L_S B J_i \tag{24}$$

the expected link bandwidth between time interval $]i, i+1]$ is

$$E[B_i^l] = (1-\lambda) E[B_{i-1}^l] + \lambda \overline{L_S} B \tag{25}$$

Then, the expected bandwidth per call is

$$E[B_S] = \frac{1-\mu}{\mu} \left(\overline{L_S} B + \lambda B_{RP} \right) \tag{26}$$

In the following section, we present the numerical results of each model and compare the results of the hierarchical handoff technique with those of the Akyol and Cox handoff scheme.

4. NUMERICAL RESULTS

The expected handoff delay in equations (8) and (21) are normalized by the signaling delay for partial rerouting D_{RP}. The expected bandwidth in equations (15) and (26) are normalized by the signaling bandwidth for partial rerouting B_{RP}.

If the time unit is second then for μ=0.03 the mean call duration is equal to 33.33 sec. The mean call duration is 1–2 minutes for the voice traffic and 15–20 minutes for data traffic, this mean durations correspond to μ0.03 and in this case, we have the following results.

Figures 5 and 6 illustrate the expected handoff delay per call as a function of λ of the hierarchical handoff and Akyol and Cox schemes. The expected handoff delay of each handoff scheme increase as λ increases. The performance comparisons between these two handoff protocols show that when D/D_{RP} increases, the hierarchical model gives an expected handoff delay lower than the one of Akyol and Cox model. This means that the higher the mobility in network, the lower the handoff delay of the hierarchical handoff scheme. When λ increases, the expected bandwidth of the hierarchical and the Akyol and Cox models increases (Fig. 7 and 8). Moreover, when B/B_{RP} is large, the expected bandwidth of the hierarchical model is lower than that of the Akyol and Cox model. This means that for networks with connections with large bandwidths the expected bandwidth per call for the hierarchical model is lower than the one of the Akyol and Cox model. We can derive a rule in which cases one would use the hierarchical or the Akyol and Cox protocols :

1 If the network operator must carry only one kind of traffic (e.g. voice) then :
   - a If the QoS (handoff delay) is more important than the bandwidth then we suggest to use the hierarchical handoff scheme.
   - b If the bandwidth is more important than the QoS then we suggest to use the Akyol and Cox handoff Scheme.

2 If the network operator must carry multimedia traffic then we suggest to use the hierarchical handoff scheme.

5. CONCLUSION

In this paper, an analytical model to evaluate the expected handoff delay per call and the expected bandwidth per call of the hierarchical handoff scheme has been proposed. This model has been adapted to the Akyol and Cox handoff scheme in order to be able to compare the expected handoff delay and the expected bandwidth per call of the Akyol and Cox and the hierarchical handoff schemes. Numerical results show that the hierarchical model yields a lower expected handoff delay and a lower expected bandwidth per call than those of Akyol and cox for the multimedia wireless ATM networks when the mobility (λ) increases.

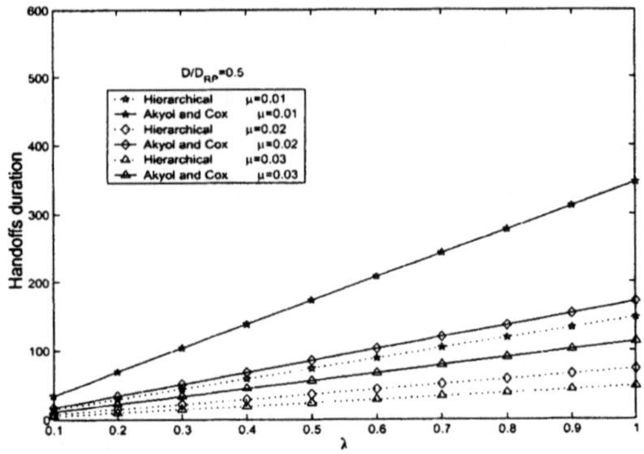

Figure 5. Expected handoff delay per call for the hierarchical and the Akyol and Cox models ($D/\ D_{RP} = 0.5,\ \overline{L_S} = 5, \overline{L_h} = 1, L_{hs} = 60$)

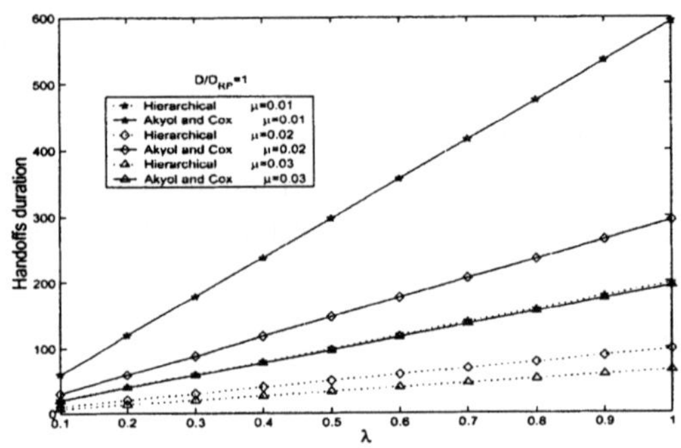

Figure 6. Expected handoff delay per call for the hierarchical and the Akyol and Cox models ($D/\ D_{RP} = 1,\ \overline{L_S} = 5, \overline{L_h} = 1, L_{hs} = 60$)

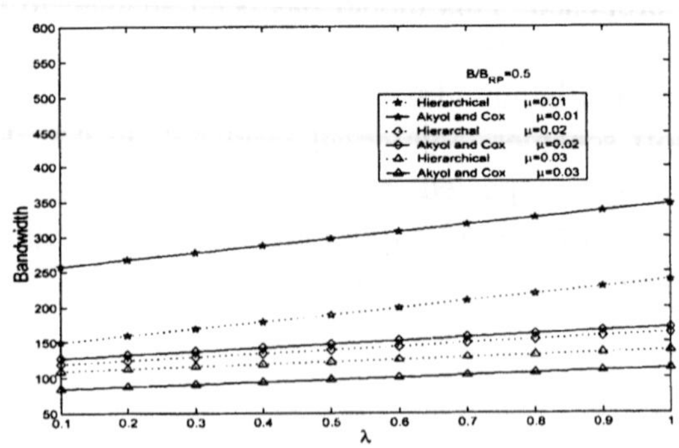

Figure 7. Expected bandwidth per call for the hierarchical and the Akyol and Cox models ($B/\ B_{RP} = 0.5,\ \overline{L_S} = 5, \overline{L_h} = 1, L_{hs} = 60$)

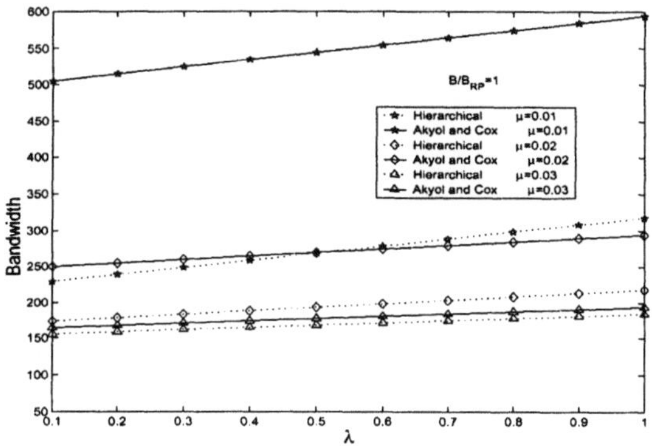

Figure 8. Expected bandwidth per call for the hierarchical and Akyol and Cox models ($B/\ B_{RP} = 1,\ \overline{L_S} = 5,\ \overline{L_h} = 1, L_{hs} = 60$)

REFERENCES

[1] Acampora and Naghshineh, 1994 Acampora, A. and Naghshineh, M. (1994). An archittecture and methodology for mobile-executed handoff in cellular ATM networks. *IEEE Journal on Selected Areas in Communications*, 12(8):1365–1375.

[2] Akyol and Cox, 1997 Akyol, B. and Cox, D. (1997). Signaling alternatives in a wireless ATM network. *IEEE Journal on Selected Areas in Communications*, 15(1):35–49.

[3] Bush et al., 1997 Bush, S., Jagannath, S., Evans, J., Frost, V., Sanchez, R., Minden, G., and Shanmugan, K. (1997). A control and management network for wireless ATM systems. *Wireless Networks*, 3:267–283.

[4] Chan and Leung, 1998 Chan, W. W. H. and Leung, V. (1998). Performance evaluations of path optimization schemes for inter-switch handoffs in Wireless ATM networks. *Proceedings of ACM/IEEE MobiCom'98, Dallas, Texas*, pages 242–251.

[5] Kara et al., 1999 Kara, N., Conan, J., and Girard, A. (1999). Handoff management in wireless ATM networks. *The Eleventh International Conference on Wireless Communications, Calgary*, pages 48–54.

[6] Raychaudhuri and Wilson, 1994 Raychaudhuri, D. and Wilson, N. (1994). ATM-based transport architecture for multiservices wireless personal communication networks. *IEEE Journal on Selected Areas in Communications*, 12(8):140–1414.

[7] The ATM Forum Wireless ATM Working Group, 1998 The ATM Forum Wireless ATM Working Group (1998). Baseline Text for Wireless ATM Specifications. *BTD-WATM-01.07.*

[8] Veeraraghavan et al., 1996 Veeraraghavan, M., Karol, M., and Eng, K. (1996). A combined handoff scheme for mobile ATM networks. *ATM Forum Contribution 96-1700.*

[9] Wong and Leung, 1998 Wong, W. and Leung, V. (1998). A path optimization signaling protocol for inter-switch handoff in wireless ATM networks. *Proceedings of* 1st *International Workshop on Wireless Mobile ATM Implementation (WMATM'98), Hangzhou, China*, pages 92–98.

Chapter 2

SD+: Improving TCP Performance over ATM-UBR Service Using Selective Drop Buffer Management Scheme

Aly E. Elabd and Mohamed A. Mostafa
Computer Engineering Department
Arab Academy for Science & Technology
Alexandria, Egypt

Key words: TCP over ATM, UBR, buffer management schemes, ATM traffic modelling

Abstract: A new packet discard scheme called Selective Drop with forgiveness mechanism that can be plugged into the existing Selective Drop buffer management scheme is developed. The new scheme that we call SD+ remarkably enhanced both throughput and fairness of TCP running over ATM-UBR service. Our scheme does not require per VC accounting but per VC-state where only one bit is needed. The focus here will be on demonstrating and analysing the effect of the forgiveness mechanism when plugged into the SD scheme in low latency networks. The SD scheme is selected because it is evolved as an improvement over both PPD and EPD policies that can be practically applied to existing ATM switches. This paper explains the concepts behind the new scheme and demonstrates its effect on TCP performance using simulation.

1. INTRODUCTION

ATM networks provide 6 different service categories: Constant Bit Rate (CBR), Variable Bit Rate (VBR-rt & VBR-nrt), Available Bit Rate (ABR), Guaranteed Frame rate (GFR) and Unspecified Bit Rate (UBR). UBR service is characterised by its simplicity, low overhead and cost effectiveness. It is also characterised by the lack of built in congestion control mechanisms [5]. The ATM FORUM traffic management specification version 4.0 states that congestion control for UBR may be performed at a higher layer on an end-to-end basis. As 85% of computer networks are running TCP/IP [13], it is expected that many TCP implementations will use the UBR service category.

TCP employs a window based end-to-end congestion control mechanism to recover from segment loss and avoid congestion collapse thus providing a reliable and adaptive connection oriented service.

Several studies have analysed the performance of TCP over the ATM-UBR service concluding that TCP applications running over ATM-UBR switches with limited buffer size suffer from low throughput and are deprived from fairly sharing the available bandwidth [1, 7, 9,13]. Accordingly, several end system side and network side enhancements had been proposed to improve this situation.

Network side enhancements focused on improving the packet discard polices of ATM-UBR switches, aiming to avoid -as much as possible-hurting TCP throughput under network congestion conditions.

Network side drop policies can be roughly divided into two categories: simple discard policies and sophisticated buffer management policies. Simple discard policies like Partial Packet Discard (PPD) or Early Packet Discard (EPD) lead to enhanced throughput but low fairness. Sophisticated mechanisms like Selective Drop (SD), Random Early Detection (RED) and Fair Buffer Allocation (FBA) improve fairness and throughput [1,3] but require per VC accounting with considerable overhead and computations that increase algorithm complexity, in addition to being parameter sensitive.

In this paper, we propose the forgiveness mechanism that can be plugged into the existing buffer management schemes to remarkably enhance both throughput and fairness. Our mechanism does not require per VC accounting but per VC-state where only one bit is needed per VC. In a previous paper [15] we demonstrated the positive effect of our proposed forgiveness mechanism over EPD. Our focus her is demonstrating and analysing the effect of the forgiveness mechanism when plugged into the Selective Drop scheme in low latency networks to enhance further SD performance.

The paper is divided into eight sections; the first one is this introduction. Section two discusses the congestion control mechanisms in the TCP protocol. The third section describes the Selective Drop mechanism while section four presents our proposed forgiveness mechanism and describes the concepts behind it. Section five describes the simulation set-up used in all our experiments and defines the performance metrics. Section six illustrates the simulation results and section seven presents the paper conclusion. Finally, section eight is a summary and ideas for future work.

2. TCP CONGESTION CONTROL

TCP uses a window-based protocol for flow control that is enforced by two windows. The sender maintains a variable congestion window (CWND) as a measure of network capacity. The receiver maintains a receiver window (RCVWND) as a measure of its receiving buffer capacity. The number of bytes that the sender can send at a time is the minimum of the two windows.

The basic TCP congestion control scheme consists of the "slow start" and "congestion avoidance" phases as shown in figure (1). The variable SSTHRESH is maintained at sender side to distinguish between the two phases and is preliminarily set to 64 KB.

When a connection is established, the sender initialises the congestion window to the size of the maximum segment in use on the connection and sends one maximum segment. If the segment is acknowledged before a certain timer expires (retransmission timer), it adds a segment worth of bytes to the congestion window to make it two segments and then sends two segments. As each of these segments is acknowledged, the congestion window is increased by one maximum segment size. In effect, each burst successfully acknowledged doubles the congestion window. The congestion window keeps on growing exponentially until either the retransmission timer expires or the receiver's window is reached; this stage is called the slow start stage of TCP.

When a time out takes place (presumingly due to network congestion), SSTHRESH is set to half the size of the current CWND and CWND is set to one segment worth of bytes, a new slow start phase is starting now. When CWND is equal to SSTHRESH, the congestion avoidance phase is started in which CWND increases by one segment every round trip time; this results in a linear increase of CWND. Figure (1) illustrates TCP congestion control mechanism [1,11,13].

It is worth mentioning that the above description covers the standard TCP described in RFC 793. However, several enhancements have been proposed to avoid TCP congestion scheme described above. We choose the standard TCP implementation as it is still the most commonly used today.

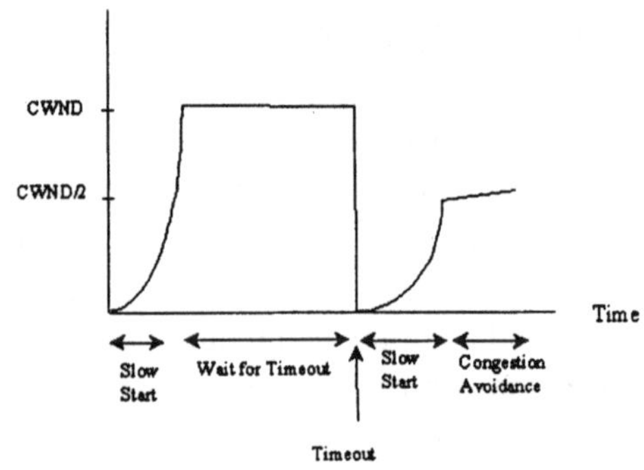

Figure 1. TCP congestion control

3. THE SELECTIVE DISCARD SCHEME

Selective Discard (SD) proposed by Goyal et Jain [1] is a buffer management scheme that can be classified among multiple accounting - single threshold (MA-ST) buffer management schemes [12]. SD evolved as an improvement over PPD and EPD policies.

PPD policy requires the switch to drop all subsequent cells from a packet if one cell has been dropped. EPD assigns a threshold smaller than buffer size, below which no cell discard takes place. If the queue length exceeds this threshold, EPD drops all arriving cells that belong to new packets while cells belonging to partially received packets are still accepted [9]. Both PPD and EPD do not guarantee fair sharing of available bandwidth between contending TCP sources [1,13].

SD ensures fair resources sharing by monitoring the share of each VC of the switch buffer. It deterministically drops cells belonging to greedy VCs in case buffer occupancy exceeds a pre-determined threshold R (similar to EPD threshold).SD keeps track of the activity of each VC by counting the number of cells from each VC in the buffer. A fair allocation is calculated as the (current buffer occupancy) divided by the (number of active VC's), where an active VC is defined as the VC with at least one cell in the buffer.

Let the buffer occupancy be denoted by X, and the number of active VCs be denoted by Na Then:

$$\text{Fair allocation} = X / Na \quad (1)$$

The ratio of the number of cells of a VC in the buffer to the fair allocation gives a measure of how much the VC is overloading the buffer i.e., by what ratio it exceeds the fair allocation. Let Yi be the number of cells from VCi in the buffer, then the Load Ratio of VCi is defined as:

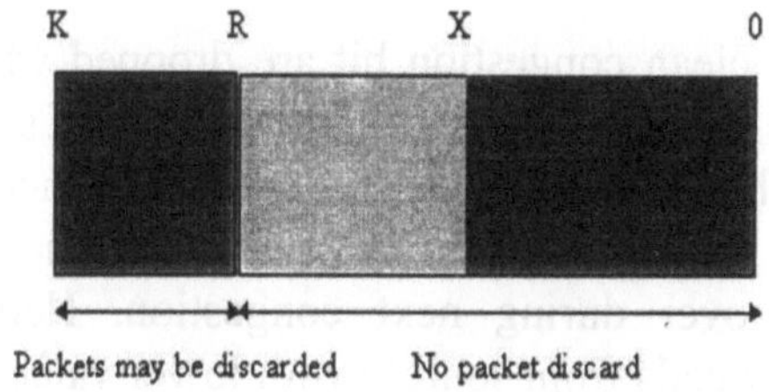

Figure 2. Selective Drop discard policy

Load Ratio of VCi = (Number of Cells from VCi) / (Fair allocation) = Yi*Na / X (2)

If buffer occupancy exceeds the static threshold R, and the load ratio of some VC is greater than a parameter Z, new packets from that VC are dropped, in preference to packets of another VC with load ratio less than Z. Thus Z is used as a cut off for the load ratio to indicate that a VC is overloading the switch. Figure (2) illustrates the buffer management policy of SD.

Goyal et Jain [1] carried an extensive set of simulations on SD concluding that the scheme is parameter sensitive (i.e. the achieved efficiency and fairness vary significantly with varying the values of Z and R). They proposed values of *Z= 0.8* and *R= 0.9 * buffer capacity* as the values that achieve the best results for efficiency and fairness.

4. SD+: THE FORGIVENESS MECHANISM

The forgiveness mechanism exploits a scheduling policy commonly used in operating systems. The scheduling algorithm tends to lower the priority of resources hungry processes in favour of lightly weighted processes. After a certain number of successive priority reductions, the scheduling algorithm forgets the past history of the heavy weighted process and treats it as a newly arriving job giving it the highest priority and a new chance to live [14].

We adapted the above concept to the ATM environment by assigning a single bit that we call the congestion bit to each VC. Whenever the congestion condition is met in SD switches (i.e. the buffer occupancy exceeds the static threshold R), newly arriving packets are continuously checked to see if they belong to greedy VCs (i.e. VCs that their buffer occupancy ratio exceeds the value of the parameter Z). If the arriving - packets belong to a greedy VC, they will be dropped and the congestion bit of its VC is set to one. In the next congestion episode, newly arriving packets belonging to VCs whose their congestion bit is set are not dropped, even if their share of buffer space exceeds the value of the parameter Z. This time greedy VCs with clean congestion bit are dropped while greedy VCs with a set congestion bit are spared. The accepted VCs will have their congestion bit reset while the dropped VC's will have their congestion bit set. Thus forgiveness mechanism has no role during the first congestion episode but it takes over during next congestion. Hence, forgiveness mechanism accepts new cells from sources with lower value of CWND even when congestion condition is met.

Floyd et Jacobson [10] state that window flow control protocols have a periodic cycle equal to the connection round trip time, and that this periodicity can resonate (i.e. have a strong non-linear interaction) with deterministic control algorithms in network gateways. They also show that drop tail gateways in a TCP/IP network with strongly periodic traffic can result in systematic discrimination against some connections. Goyal et Jain [1] demonstrate the effect of such periodicity on TCP throughput and fairness in ATM networks with tail drop and early packet discard drop policies.

The forgiveness mechanism effectively helps reducing TCP synchronisation effects described in [1] and [10] and helps promoting randomness element in the network, hence breaks down phase effects and diminishes resonance responsible for efficiency degradation. This is accomplished by selectively and deterministically dropping connections with larger CWND while keeping other connections untouched in one congestion episode and reversing this action in the next one, the thing that will also enable TCP sources to exploit a fair share of the available network capacity in turn.

It is clear from section 2 that successive congestion badly hurt TCP efficiency. As the size of SSTHRESH will dramatically decrease leading to a situation in which the slow start phase, where CWND size increases exponentially, will be too short increasing the time taken by the TCP source to reach steady state after congestion breakdown. The forgiveness mechanism remarkably increases throughput of TCP sources as it prevents successive reduction of SSTHRESH.

The forgiveness mechanism requires minimal housekeeping overhead (one bit per VC). Also the logical condition it enforces can be easily applied in hardware which makes it compatible with ultra fast switching requirements of ATM networks and easily applicable in existing ATM switches.

As stated earlier the forgiveness mechanism can be plugged into several packet discard mechanisms to enhance the efficiency of TCP sources. We will show in this paper such effect in networks that apply SD switch buffer management scheme.

To evaluate the effect of the forgiveness mechanism, we use simulation, an approach commonly used in many other papers that dealt with the subject of traffic management of TCP/IP over ATM networks [1,2,3,7,9]. We use the ATM/HFC network simulator version 4.1, released December 1998, by the National Institute for Standards and Technology –NIST (USA). The simulator source code has been modified to add the forgiveness mechanism condition to the SD drop policy. Simulation parameters and performance metrics are described in details in the following section.

5. SIMULATION MODEL, PARAMETERS AND PERFORMANCE METRICS

5.1 Simulation model

All simulations presented in this paper use the N source configuration shown in figure (3). This model has been widely used in other papers discussing performance issues of TCP over ATM [1,2,3,7]. The model consists of N identical TCP sources that share a single bottleneck. The switches used in this model implement UBR service with either SD or SD+ (SD with the forgiveness mechanism plugged in).

5.2 Simulation parameters

Selection of parameters is in accordance with the guidelines in [8]. The following simulation parameters are used:

- The configuration consists of N identical TCP sources as shown in figure (3).
- All sources are ftp sources with very large file sizes (30 M B); ftp is a very common traffic pattern, large file sizes ensure continuos data flow throughout the simulation period.
- ATM applications are assumed to be running over SONET STS-3c links with bandwidth equal to 155.52 Mbps.
- The following link lengths have been tested: 0.8 Km, 1 Km, 2 Km and 3 Km per link.
- Switches are assumed to implement SD/SD+ discard policies; R parameter is set to *buffer size* 0.90* cells and Z parameter is set to 0.8;these values are selected because they ensure optimum results for efficiency and fairness as described in [1].
- Switch buffer sizes of 1000 and 3000 cells have been tested respectively.

- Peak cell rate is 146.9 Mbps (after factoring in SONET overhead).

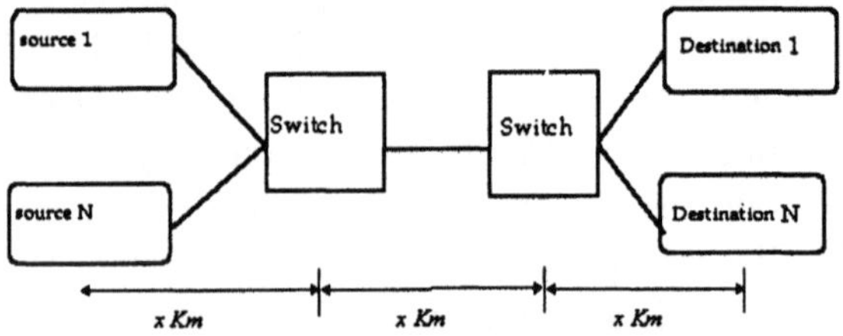

Figure 3. N-Source configuration

- Traffic is unidirectional, only sources can send data and destinations send acknowledgements only.
- TCP maximum segment size is set to 512 bytes; this is the standard TCP segment size. Testing with larger segment sizes is left to future work.
- TCP retransmission timer granularity is set to 100 ms.
- TCP RCVWND is set to 64 KB; this is the default TCP receiver window size.
- All TCP sources start sending data at the same time and keep on sending data throughout the simulation, this pattern increases the probability of TCP synchronisation.
- Simulated duration is 10 seconds.

5.3 Performance metrics

Two performance metrics are used: efficiency and fairness, the same set of metrics have been used in several studies for performance measurements of TCP over UBR [1,3,7]. They also conform to the recommendations of the test-working group of the ATM FORUM [6].

Efficiency can be defined as the ratio of the goodput achieved by all TCP

$$\sum_{i=1}^{i=N} \frac{Xi}{C} \qquad (3)$$

connections to the maximum achievable throughput of TCP over ATM-UBR, interpreting the above sentence to a mathematical formula, we get:
Where N is the total number of TCP sources, x_i is the goodput of TCP source number i and C is the maximum achievable throughput of TCP over a certain ATM link.

The goodput of a TCP source is defined as the total number of successfully transmitted bytes (excluding retransmissions) divided by the

simulation time. The maximum possible TCP throughput over UBR can be calculated as follows:
For a 512 bytes TCP segment size, ATM layer receives the following payload:

8 byte LLC header	20 byte IP header	20 byte TCP header	512 byte TCP payload	8 byte AAL5 trailer

Figure 4. Payload at the ATM layer

This payload is padded to form 12 ATM cells, thus 512 data bytes at the TCP layer results into 636 bytes at the ATM layer.

So, the maximum possible throughput is 512/636 =80.5 %=125.2 Mbps on a 155.52 Mbps link. For ATM over SONET, this number is further reduced to 120.5 Mbps [1,11]. Fairness is measured by the fairness index F defined by:

$$FairnessIn\ dex(F) = \frac{\left(\sum_{i=1}^{i=N} \frac{xi}{ei}\right)^2}{N * \sum_{i=1}^{i=N} \left(\frac{xi}{ei}\right)^2} \quad (4)$$

Where e_i is the fair share of source i of the available bandwidth for the N source configuration $e_i = C / N$. A fairness value of 0.9 may or may not be accepted according to the nature of application and the number of sources involved. A fairness index of 0.99 is considered to be near perfect [1].

6. SIMULATION RESULTS

We simulated 5 TCP sources with finite buffer switches. The simulations were performed using two values of buffer sizes 1000 and 3000 cells respectively, these values were chosen because they are closer to existing LAN switches and have been used in other studies [1]. The results are illustrated in the following tables:

Table 1. SD Vs. SD+ (Efficiency)

Buffer size (cells)	Link length(Km)	SD	SD+
1000	0.8	0.55	0.62
3000	***0.8***	***0.67***	***0.89***
1000	1	0.51	0.6
3000	***1***	***0.71***	***0.91***
1000	2	0.66	0.62
3000	***2***	***0.69***	***0.89***
1000	3	0.35	0.6
3000	***3***	***0.71***	***0.91***

Table 2. SD Vs. SD+ (Fairness)

Buffer size (cells)	Link length(Km)	SD	SD+
1000	0.8	0.98	1
3000	0.8	***0.97***	***0.99***
1000	1	0.87	0.99
3000	1	***0.99***	***0.99***
1000	2	0.99	0.98
3000	2	***0.97***	***0.99***
1000	3	0.85	0.94
3000	3	***0.97***	***0.99***

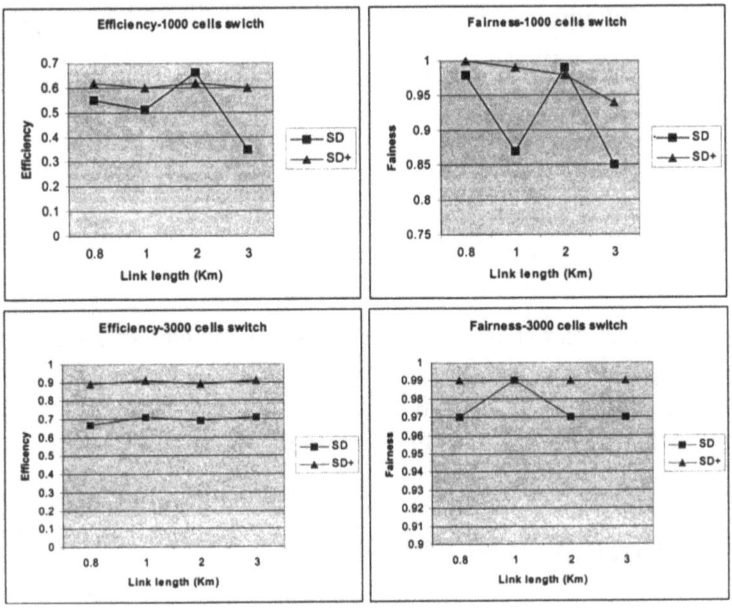

Figure 5. Performance of SD+ Vs. SD

Now, we want to examine the performance of SD+ in case of large number of sources; we simulated 10 TCP sources with finite buffer switches. The simulations were performed using two values of buffer sizes 2000 and 3000 cells respectively for link length of 0.8 Km, 1 Km and 1.5 Km; simulation results are illustrated below.

Table 3. SD+ Vs. SD for large number of sources (Efficiency)

Buffer size (cells)	Link length(Km)	SD	SD+
2000	0.8	0.73	0.93
3000	0.8	***0.32***	***0.93***
2000	1	0.54	0.94
3000	1	***0.93***	***0.94***
2000	1.5	0.60	0.89
3000	1.5	***0.96***	***0.93***

Table 4. SD+ Vs. SD for large number of sources (Fairness)

Buffer size (cells)	Link length(Km)	SD	SD+
2000	0.8	0.91	0.98
3000	0.8	***0.73***	***0.97***
2000	1	0.87	0.98
3000	1	***0.96***	***0.97***
2000	1.5	0.92	0.98
3000	1.5	***0.94***	***0.96***

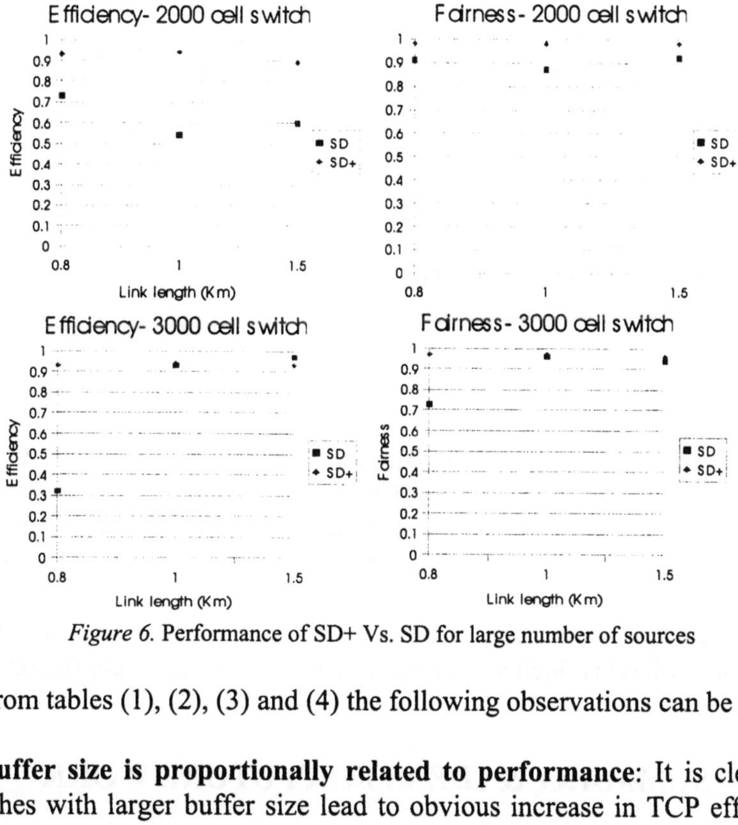

Figure 6. Performance of SD+ Vs. SD for large number of sources

From tables (1), (2), (3) and (4) the following observations can be made:

Buffer size is proportionally related to performance: It is clear that switches with larger buffer size lead to obvious increase in TCP efficiency and configuration fairness; under same network conditions. The switch with 3000 cells buffer capacity leads to 10% ~ 40% rise in the efficiency and 1% ~ 7% rise in fairness index than that of 2000 cells buffer size.

SD+ remarkably enhances performance of TCP over ATM-UBR: Remarkable increase in efficiency is observed when applying SD+. This improvement is due to SD+ drop policy that breaks synchronisation between TCP sources. Hence decreasing the probability of switch 's buffer overflow which negatively affect TCP throughput as it forces TCP sources to enter the slow start phase decreasing CWND to 1 as described in section 2. SD+ most remarkable efficiency improvement is achieved in the case of 2000 cell switches network where about 48% increase in efficiency is observed while the highest efficiency value (0.94) is achieved in the network with 3000 cell buffer switches.

SD+ enhances fairness: SD+ increases efficiency but does not sacrifice fairness, fairness value achieved with SD+ is higher than that obtained when applying SD in most cases. Even when SD achieves a suitable fairness index (0.99) in networks with larger switch buffer sizes, it achieves the same value or even increases it to the ideal value (1). This is due to SD+ drop policy which fairly allocates the bandwidth between the contending sources by granting the available network resources to TCP sources with lower CWND.

SD+ has a steady state performance: It is clear from simulation results that SD+ has a steady state performance in all configurations simulated where the values of achieved efficiency and fairness are almost constant, while the performance of SD was fluctuating between good performance and unexpected performance breakdown.

7. CONCLUSION

SD+ enhances efficiency and fairness over SD. The positive effect is obvious in switches with lower buffer sizes (1000 & 2000 cell switches) where remarkable improvement is observed. In case of switches with 3000 cells buffer size, the achieved values of efficiency and fairness are almost equal to the values achieved by SD.

SD+ performance is stable in different network configurations while the performance of SD is fluctuating when tested under the same configurations.

8. SUMMARY & IDEAS FOR FUTURE WORK

TCP traffic phase effects and poor buffer management policies lead to TCP performance degradation in ATM-UBR networks. Simple drop policies like EPD considerably improve efficiency but it is not guaranteed to provide fairness. Sophisticated drop mechanisms like SD and FBA improve both efficiency and fairness much further using per VC accounting that require considerable accounting overhead and increased algorithm complexity, in addition to being parameters sensitive. Our proposed forgiveness mechanism can be used as a plug in to any of the above mentioned schemes to remarkably improve performance and fairness, we showed how the forgiveness mechanism can improve the performance of SD yet requiring minimal overhead (1 bit per VC).

In this paper, we focused on analysing the effect of forgiveness mechanism over SD. In a future paper we will show how the forgiveness mechanism can considerably improve the performance of FBA, it is also interesting to investigate the effect of larger number of TCP sources and larger TCP segment sizes on TCP throughput when using SD+.

REFERENCES

[1] Rohit Goyal, Raj Jain, Shiv Kalyanaraman, Sonia Fahmy and Boby Vandalore, "Improving the performance of TCP over the ATM-UBR service," Journal of Computer Communications Vol. 21, No. 10, July 1998.

[2] Shivkumar Kalyanaraman, Raj Jain, Sonia Fahmy and Rohit Goyal, "Performance and buffering requirements of Internet protocols over ATM ABR and UBR services", IEEE communications Magazine,June 1998

[3] Vincent Rosolen, Olivier Bonavnture and Guy Leduc, "Impact of cell discard strategies on TCP/IP in ATM UBR networks", Proc. of the 6th Workshop on Performance Modeling and Evaluation of ATM Networks (IFIP ATM'98) Ilkley, UK,July 98.

[4] Kenji Kawahara, Koichiro Kitajima, Ttetsuya Takine and Yuji Oie, "Packet loss Performance of Selective Cell Discard Schemes in ATM Switches", IEEE Journal on selected areas in communications, Vol. 15, NO. 5, July 1997.

[5] ATM FORUM, ATM Traffic Management Specification Version 4.0, Apr. 1996.

[6] Raj Jain and Gojko Babic, "Performance testing effort at the ATM FORUM : An Overview", IEEE Communications Magazine August, 1997.

[7] Allyn Ramanov, Sally Floyd, "Dynamics of TCP traffic over ATM Networks", IEEE Journal on Selected Areas In Communications, May , 1995.

[8] Tim Dwight, "Guidelines for the simulation of TCP/IP over ATM", ATM FORUM 95-0077r1, March 1995.

[9] Raj Jain, Rohit Goyal, Shiv Kalyanaraman, Sonia Fahmy and Fang Lu, "TCP/IP over UBR", ATM FORUM/96-0179.

[10] Sally Floyd and Van Jacobson, "On Traffic Phase Effects in Packet-Switched Gateways", Computer Communication Review,V.21N.2, April,1991.

[11] Andrew S. Tanenbaum. *Computer Networks: Third Edition, Prentic* -Hall, Inc., 1996.

[12] Rohit Goyal, "Traffic management for TCP/IP over Asynchronous Transfer Mode (ATM) Networks", Ph.D. Dissertation, Ohio-State University, 1999.

[13] Xiangrong CAI, "The performance of TCP Over ATM ABR and UBR Services", http"//www.cis.ohio-state.edu/~jain/cis788-97/tcp_over_atm/index.htm

[14] Harvey M. Deitel, *An Introduction to Operating systems: second edition*, Addision-Wesley publishing company, 1990.

[15] Aly E. El-Abd and Mohamed A. Mostafa, "EPD+: A New Packet Discard Mechanism to Improve TCP Performance over ATM-UBR", proceedings of ICATM2000.

REFERENCES

Chapter 3

Performance Analyses of Access Points In Cellular Wireless ATM Networks

A Marshall, F Garcia-Palacios, S Sezer, D Chieng
School of Electrical and Electronic Engineering, The Queen's University of Belfast, Belfast BT9 5 AH, Northern Ireland, U.K

Key words: Quality of Service, Mobility, Wireless ATM, Access Point.

Abstract: The performance of Wireless Asynchronous Transfer Mode (WATM) networks is influenced by both cell level and call level issues. These can introduce delay and loss into the services that use the WATM Access Points (APs). At the cell or burst levels, buffering and schemes in the APs, and the instantaneous load applied to the radiocells can have an impact on the Quality of Sevice (QoS) offerred. At the call level, the number of calls blocked at setup or handoff contributes to the Grade of Service (GoS) offered. The GoS and QoS must be considered together when assessing the performance of multimedia services in a WATM network. A novel modelling approach is presented here, whereby an integrated call level-cell level model allows realistic traffic patterns to be generated and applied at both levels. This approach permits analysis of a WATM AP architecture in terms of delay (at the cell level), and blocked calls (at the call level). Results presented show the performance of different buffering techniques in the WATM APs. Additionally, call level performance parameters allow the GoS offered to clients in the radio picocells to be adjusted across a range of traffic classes in order to optimise the QoS for time-sensitive applications.

1. INTRODUCTION

Wireless ATM (WATM) is an extension of fixed ATM networks, and provides mobile users with wireless access to broadband multimedia services, similar to those expected to be provided by fixed access ATM networks. Hence, WATM has to provide high bandwidth and support many services with different QoS objectives, and with traffic profiles similar to those encountered in fixed ATM networks. The importance of QoS has been recognised in recent years as critical for achieving predictable B-IDSN network performance. The provision and support for QoS was defined within the early standardisation of ATM [6] and its detail specified by the ATM-Forum [1], [2].

The poor transmission link quality in WATM, requires a much greater effort in order to achieve the target QoS for all connections. Besides the Physical radio layer, the ATM and the MAC layers may be viewed as two layers that also influence QoS. In the MAC layer, QoS guarantees may be violated by a poorly designed access scheme [7]. In the ATM layer one of the most important issues that impact QoS is buffer performance in the AP [13].

At the system level, call parameters such as arrival rate and session length can have a direct influence on the dynamic load at the AP (i.e. they dictate the dynamic load presented to the buffers). Traditional call level performance parameters such as the call blocking probability also give an idea of how well the system performs (i.e. the call level performance). This forms part of what may be termed the Grade of Service (GoS) provided by the WATM network. Current research considers both fields independently [9,14], while this paper considers that they are very much interrelated.

This paper describes the development of a novel approach for modelling WATM, which employs integrated call and cell level models of a tessellated picocellular WATM access network. The network supports four classes of multimedia traffic, Constant Bit Rate (CBR), Real-time Variable Bit Rate (rt-VBR), Non Real-time Variable Bit Rate (nrt-VBR) and Unspecified Bit Rate (UBR). These are termed Class 1, Class 2, Class 3 and Class 4 services respectively throughout this paper. The approach allows analysis of the dynamically changing, and multi-class traffic loads presented to the AP. The call level performance of the network is characterised by several call parameters, while the cell level analysis allows the evaluation of a number of buffering strategies considered for use in the APs.

2. PERFORMANCE OF A WATM SYSTEM

For a multi-class traffic network, two parameters must be considered: QoS, which concerns cell loss and delay, and GoS, defined here as the ratio between call time granted and call time requested for each traffic class. GoS and QoS should be considered together. If the system achieves very low blocking probabilities (i.e. high GoS) but the buffers are overloaded, then connections may suffer from considerable delay or cell loss (i.e. QoS not achieved). Hence there is a compromise between connections accepted and connections delivered with QoS.

At connection establishment, the network management must estimate the network resources required. It is necessary to provide information about the multimedia traffic characteristics, required bandwidth and QoS requirements, in order to guarantee a desired level of service. The ATM Forum has specified the traffic contract parameters (Traffic Descriptors) [1]. Negotiation of a connection is provided through the Connection Admission Control (CAC) in conjunction with a Wireless CAC [9] (WCAC) that manages the radio resources in WATM. A connection is progressed only when sufficient resources are available at each successive network element to establish the connection through the whole network, based on its service category, traffic contract, and QoS; and in order to maintain the agreed QoS for existing connections [3, 1, 6].

At connection establishment, the network management must estimate the network resources required. It is necessary to provide information about the multimedia traffic characteristics, required bandwidth and QoS requirements, in order to guarantee a desired level of service. The ATM Forum has specified the traffic contract parameters (Traffic Descriptors) [1]. Negotiation of a connection is provided through the Connection Admission Control (CAC) in conjunction with a Wireless CAC [9] (WCAC) that manages the radio resources in WATM. A connection is progressed only when sufficient resources are available at each successive network element to establish the connection through the whole network, based on its service category, traffic contract, and QoS; and in order to maintain the agreed QoS for existing connections [3, 1, 6].

From the traffic descriptors and QoS requirements, the CAC and WCAC estimate the amount of bandwidth needed to achieve the required QoS for an individual connection. There are many bandwidth estimation algorithms [3]. Allocation of bandwidth in our model is based on the Sustainable Cell Rate (SCR), and on an estimated rate midway between the SCR and the Peak Cell Rate (PCR), termed Modified Cell Rate (MfCR) [8]. This is obtained by:

$$MfCR = SCR + A\,(PCR - SCR) \quad (1)$$

Cell loss in the buffer is not considered here, however this has only a minor effect on QoS since large buffers are provided (100K cells) as is the trend in commercial switches [10]. The analysis presented here is mainly concerned with delay sensitive traffic. Cell delay and hence QoS can be improved by implementing different queuing strategies or by employing MfCR in the schedulers for delay sensitive connections.

At the call level, the parameters which influence GoS are: the number of calls serviced (normal calls), the number of blocked calls (at set up and handoff) and the "out of coverage area" calls. By minimising the number of calls blocked, the GoS will be improved. Minimisation of calls blocked at handoff is considered a critical issue, since more connections can proceed without their session being interrupted. This is beneficial even if it is done at the expense of a small number of extra calls being blocked at set-up, since they can be queued and serviced later. The number of handoffs performed for each traffic class is also important, since handoff implies a service disruption and by minimising this number, QoS can also be improved.

3. MODELLING AND SIMULATION

In order to undertake the analysis, a computer-based Mobile ATM Call level-Cell level system has been developed [5]. In this model various buffering schemes can be implemented at the AP. Call level performance parameters can also be measured individually for each traffic class. To achieve a realistic scenario, every mobile connection has an individual traffic source, which is modelled according to the class of the connection. Mobile terminals are created randomly in the coverage area and given call level parameters such as session length and speed. For each connection, the corresponding traffic source is activated at the cell level.

The call level parameters can dynamically vary the load offered to the AP. These are assigned independently according to the traffic class, for example a video connection (class 2 traffic) will have very low mobility and a long session length.

3.1 TRAFFIC MODEL (Cell Level)

Four traffic classes have been considered, with the following traffic characteristics:

- Class 1.- 64 Kbps voice connections with constant inter cell delay and burst size are used for constant bit-rate (CBR) traffic.
- Class 2.- MPEG2 Video Conferencing Source according to [12] is used for real-time variable bit-rate (rt-VBR) traffic.

- Class 3.- A nrt-VBR source based on a standard Ethernet traffic source model, which statistically models traffic by generating bursts from a bimodal distribution with exponential silence times between bursts.
- Class 4.- A random traffic source with exponentially distributed inter-burst delay and geometrically distributed burst size is used as an Unspecified Bit Rate (UBR) source (e.g. File transfer and email).

The objective of these traffic models is to create the superposition of bursts from individual mobile VC's, as occurs in practice [15].

3.2 THE INTERACTIVE CALL LEVEL/ CELL LEVEL MODELLING APPROACH

At the call level, the system emulates an indoor scenario where users move randomly through a tessellated cluster of WATM picocells. The model assumes that mobile users' arrivals into the coverage area is Poisson distributed. The attributes of the user, such as speed of motion, direction, and session length, are randomly generated from appropriate uniform distributions. Each user is assumed to move along a set of straight lines in different directions (updated at certain intervals) for the duration of a session. As users arrive in a cell, they request a certain bandwidth (depending on their connection class) from the AP. If the AP has the requested bandwidth available the connection is created, otherwise the call is blocked. As the user moves, the channel quality, measured as a function of distance to the AP, is periodically checked. If the quality is not acceptable the channel is released and another AP and channel are assigned to the mobile user (i.e. hard handoff is performed).

The handoffs and blocked calls in each cell can be monitored, and this in turn can be used to dynamically adjust the cell level traffic sources so that the load presented to the buffers in the APs will also vary dynamically. The Call level model passes 4 parameters (the number of users in each class) to the cell level model. These parameters are updated at call set up, call termination and during handoffs. The corresponding traffic sources are applied to the AP buffers, allowing the performance of each type of buffering strategy to be analysed.

4. THE CALL LEVEL MODEL - RESULTS

The simulations presented here consider a tessellated 7-cluster picocellular WATM network with 10m cell radii, covering an area of approximately 5000m^2 with 16 cells. The capacity of each picocell was set to

10 Mbps. Within specific picocells mobile populations will vary due to call establishment and termination, and handoffs. Calls can be blocked at either setup or handoff. Calls are terminated because of two reasons i) the session length is over (normal call), or ii) the user goes out of coverage area. The call parameters are given in Table 1 and are similar to those considered in AWACS ACTS project [4].

Table 1. Call Level Set Up Parameters

Call level Parameters	Class 1 (CBR)	Class 2 (rt-VBR)	Class 3 (nrt-VBR)	Class 4 (UBR)
Interarrival T.(calls/s)	15	20	10	20
MaxSpeed limit(m/s)	1.5	0.1	0.5	1
MinSpeed limit(m/s)	0.01	0.005	0.005	0.01
MaxSession length(s)	240	3600	1800	60
MinSession length(s)	15	1800	300	10
Capacity mean (bps)	64000	1300000	500000	500000

Calls may be displayed according to their traffic class, call type and blocking type as shown in Table 2. In these results a connection is accepted based on the SCR of the traffic class. The results are obtained for the total coverage area and one hour call time simulation. Alternatively any cell in the coverage area can be monitored individually.

Table 2. Call Statistics for GoS

Call Type	Class 1 (CBR)	Class 2 (rt-VBR)	Class 3 (nrt-VBR)	Class 4 (UBR)	Total by Type
Normal	115	29	107	157	408
Out of Region	119	2	114	11	246
Blocked at Set-up	0	19	14	9	42
Blocked at Hand-off	0	3	37	4	44
Total by class	234	53	272	181	740

The blocking ratios can be assessed from Table 2. These are high for classes 2 and 3, due to the high bandwidth requested and the long session times of those classes. Class 4 has a moderate blocking ratio due to its short session length, and class 1 (voice connections) performed extremely well considering that out of coverage area calls, which are high due to the high mobility of these users, could be covered by deploying more APs. The total number of handoffs, and handoff per class can also be monitored. As expected from a picocellular environment, the handoff rate is high (an average of 1 handoff every 40secs in a cell), although classes 2 and 4 have much lower handoff rates due to their low mobility or short session time. These results picture a realistic scenario in a multimedia indoors environment and provide the metrics to assess the GoS at the call level. This is computed as the percentage of the total time requested to the time granted.

Figure 1 shows the effect of increasing cell capacity (bandwidth) upon the GoS. In particular, the results show that the major beneficiaries of such bandwidth prioritisation are the class 3 users (nrt-VBR) who are highly mobile. An alternative strategy entails reserving existing bandwidth in a picocell specifically for class 3 users attempting handoff (figure 2). This significantly decreases the number of class 3 calls blocked at handoff, and the results show it is beneficial if only a small amount (5%) of the bandwidth is reserved. Reserving more than 15% rapidly increases the total number of class 1 calls blocked during handoff. This suggests that bandwidth prioritisation, or reallocation from neighbouring cells, especially for highly mobile and burst users, should be studied further.

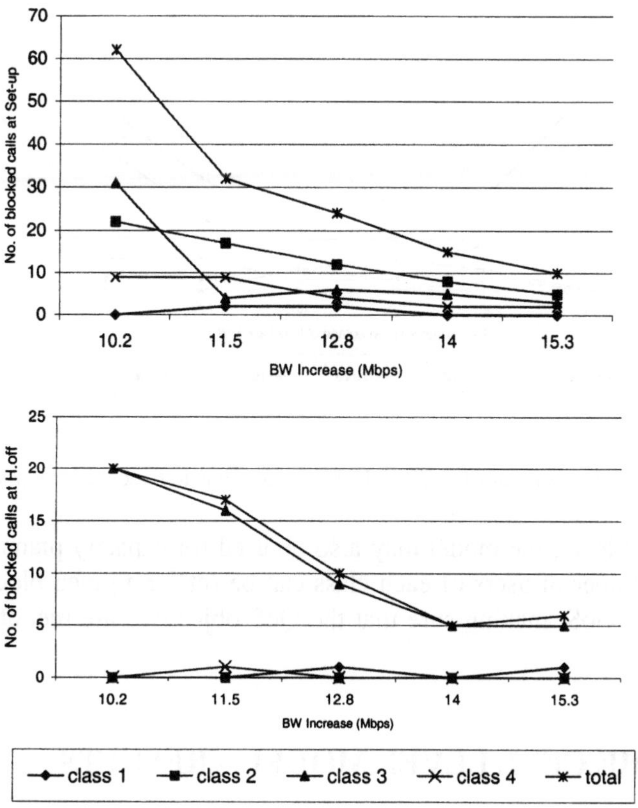

Figure 1. Decrease of blocked calls due to Cell Bandwidth increase (16 cells, r=15m)

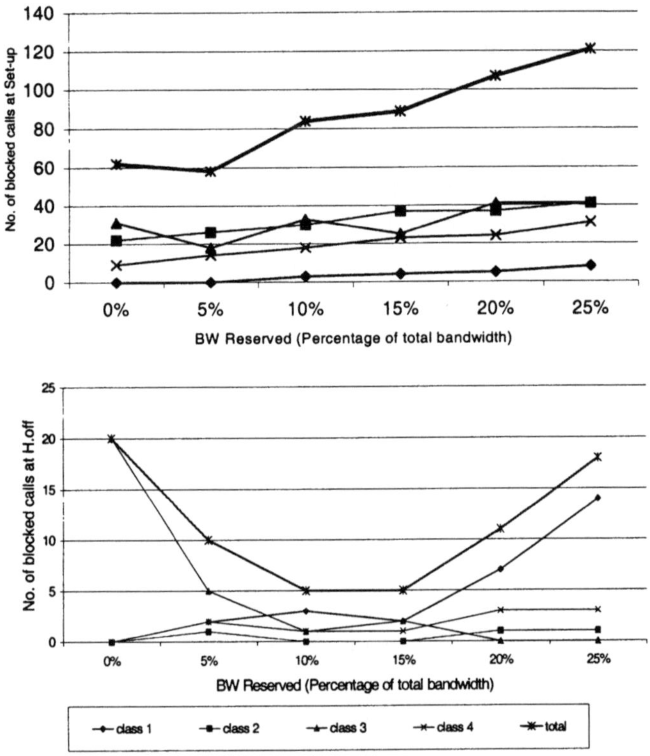

Figure 2. Decrease of Class 3 blocked at handoff due to Bandwidth reservation

At the call level, the model may also be used for capacity planning. The maximum number of users of each class can be set for a given blocked call probability or GoS, making sure that the QoS objectives are met at the cell level.

5. THE CELL LEVEL MODEL – RESULTS

At the cell level, the performance of several buffering techniques that can be used in the AP are analysed in terms of delay. The analysis considers downstream buffering, as this is the most critical path. That is, the capacity of the picocell is generally less than the capacity of the fixed link [13].

5.1 BUFFERING TECHNIQUES

Three buffering techniques are considered:-

The single queue approach: the cells arriving from the switch are queued in a FIFO manner. All traffic classes are queued into a single queue. While this model represents a very simplistic AP, it nevertheless serves as a reference against which the performances of more sophisticated models can be compared.

The traffic profile at the AP can be expected to be heterogeneous, therefore the bursty and variable characteristics of best-effort type traffic and VBR traffic can cause random cell transmission delays with associated jitter (CTD, CDV) for all of the connections at the access point. This unpredictable nature of the delay when using a single queue is the cause of QoS violation for CBR and rtVBR services [11].

Per-class queuing approach : In this approach one FIFO queue is used for each class of traffic. Connections of the same class are queued in one of the four corresponding queues. The number of cells NS_{Class}, which will be serviced for a particular class is determined by:

$$NS_{Class}(t) = \frac{B_{Class}(t) \cdot T_{SCycle}}{8 \cdot 53} \quad (2)$$

NS_{Class} : maximum number of serviced cells for a class during a scheduling cycle.

B_{Class} : Allocated bandwidth of the class in a link.

T_{SCycle} : Scheduler cycle time

If the serviced queue is empty before the allocated service time slot is finished, the next queue will be serviced, in order to utilise the remaining time-slots (i.e. work-conserving scheduling). Different scheduling algorithms may be considered to improve QoS at the AP, although they are outside the scope of this paper. B_{Class}(t) will change over time as the number of users of a particular traffic class in a picocell vary. In our simulation this is achieved by the call level model passing the number of users per class to the cell level scheduler.

Per Virtual Connection queuing approach (Per-VC): In this approach, one queue is allocated per established connection. Queue servicing is performed according to the allocated bandwidth of each connection. Fair servicing over all queues is important in order to provide predictable delay. The scheduler allocates a number of service timeslots ST_i for each connection during each service period.

$$ST_i = \delta_i \cdot T_{MSP} \quad (3)$$

T_{MSP} = The main service period with a fixed time interval
δ_i = A certain percentage of T_{MSP} for this connection.

If a connection is not in use, the timeslots can be passed to the next connection. Since every connection is allocated an individual buffer, the average delay for a particular connection i can be obtained as a function of the utilisation of the allocated bandwidth for that connection. Interference from other connections is thus avoided [11].

5.2 CELL LEVEL RESULTS

The results in figures 3 to 7 show the performance of different buffering techniques. The results were measured at the centre AP in a cluster. The call time simulated was 30 minutes and the instantaneous total load is also shown. Figures 3 and 4 compare single queue against per-class performance. The major improvement here is with CBR traffic. It may be observed that for per-class queuing the mean delay for class 1 never exceeds 1ms, compared to the single queue system where there are periods when the delay approaches 1 second. It may also be observed however, that the single queue performs better for rt-VBR traffic. The reason for this is that the per-class scheduler allocates bandwidth according to the SCR of the source. However the peak rate of the rt-VBR traffic sources is nearly 5 times larger than its mean rate. This means that whenever a large traffic burst occurs from this type of source, it cannot be completely serviced within the schedule time allocated to the rt-VBR queue. In a single queue system however, all bursts are serviced at the full line rate and rt-VBR connections benefit. Nrt-VBR and UBR traffic delays are also shown for completeness, although delay is not significant for these classes.

In Figure 5 a new bandwidth allocation policy was implemented in the per class scheduler. Bandwidth is no longer allocated according to the SCR, instead it has been increased from 1,378,000 bit/s to a MfCR of 1,929,000 bit/s for rt-VBR sources and decreased from 500,000 bit/s to a MfCR of 256,000 bit/s for non delay sensitive traffic (nrt-VBR and UBR). It can be observed from figure 5, that the results are excellent for real time services (always below 0.1 sec.). When the load is high, nrt-VBR services have high delay, which could be decreased by increasing the bandwidth allocated in the scheduler if a specific application requires it.

Figures 6 and 7 show that a per-VC scheduling according to MfCRs performs only slightly better than scheduling according to SCR at the expense of extra bandwidth. Hence, an optimal technique would be per-VC scheduling according to a SCR allocation scheme.

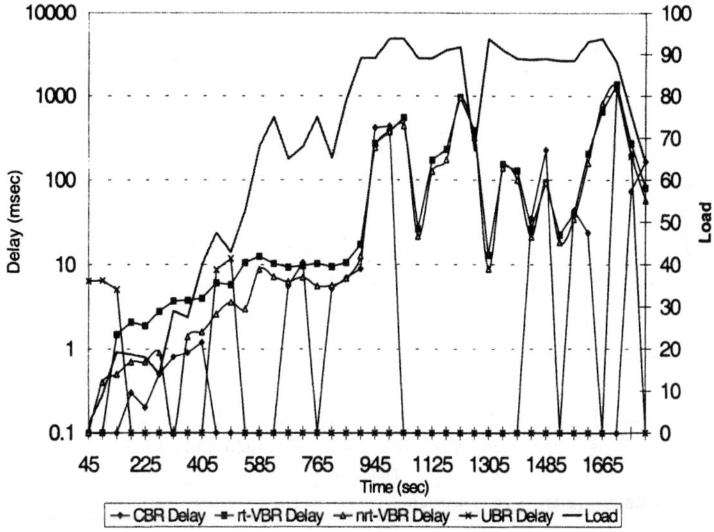

Figure 3. Mean Delay Single Queuing

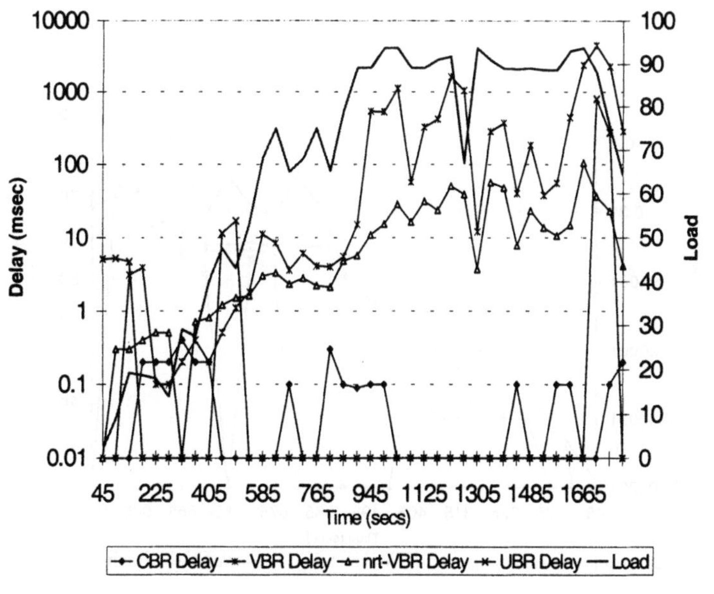

Figure 4. Mean Delay Per Class Queuing

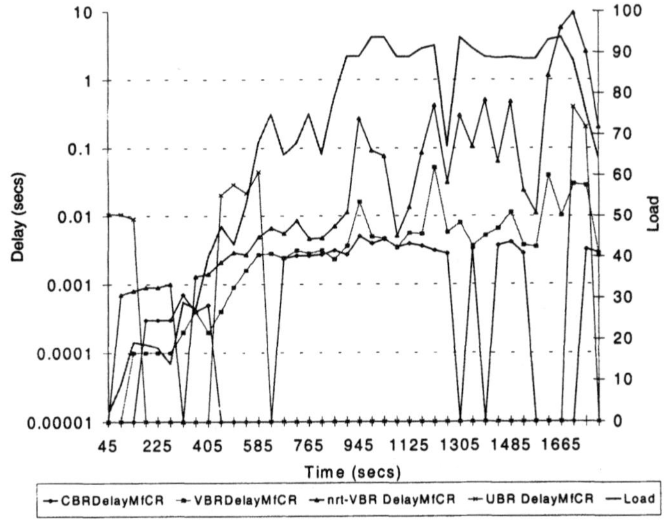

Figure 5. Mean Delay Per Class with MfCR allocation

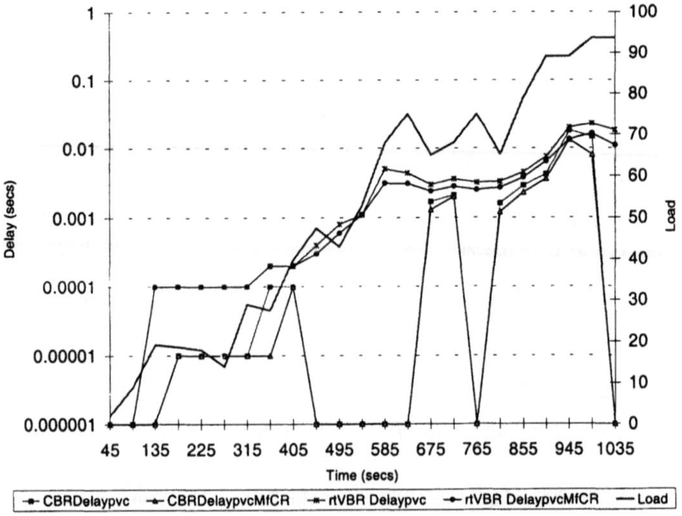

Figure 6. Mean Delay for Real Time Services (PVC with SCR and MfCR allocation)

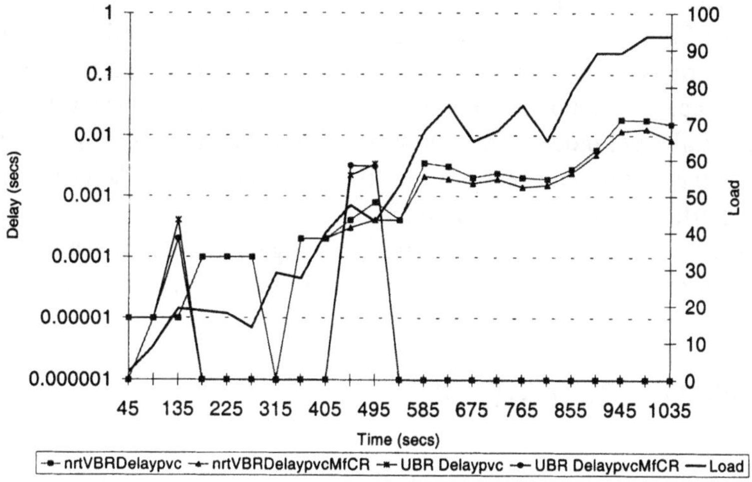

Figure 7. Mean Delay for Non-Real Time Services (PVC with SCR and MfCR allocation)

6. CONCLUSION

This paper has shown that QoS in WATM networks is affected by the delay introduced in wireless access points. The paper has analysed these delays whenever a dynamically varying load is applied in a mobile multimedia environment. A novel approach to modelling the dynamic load presented to the access point has been developed by employing a two-level simulation. A call-level model of the mobile users and their applications is used to predict the dynamically changing load per class of traffic that is offered to the access point. This in turn is used by a cell-level model to analyse the performance of any buffering and scheduling strategies employed in the access point. Simulations have been undertaken for a tessellated picocellular WATM network providing multi-class services and three buffering approaches in the access point (single, per-class and per-VC queuing) have been examined.

The analysis shows that the access point buffer delay, is influenced by each connection when using a single queue. This causes unpredictable delays for all traffic when such a queue is used. This causes QoS violations for CBR and rtVBR services. An improvement is obtained by applying per-class and per-VC buffering. Also, scheduling according to rates above the SCR can improve delay significantly in a per class scheme. However per-VC scheduling above the SCR yield only slight improvement. The extra complexity required to implement this, may be traded against MfCR per-class scheduling, which is easier to implement and produces similar performance.

The call level – cell level model interaction is also very important, in the sense that for a given call blocking probability (GoS) the model can assess if QoS (in terms of delay) at the cell level is achieved. The effect of increasing picocell capacity (bandwidth) on the GoS for each traffic class was analysed with the call level model. This shows rt-VBR is the main beneficiary, which suggests that bandwidth reallocation from neighbouring cells, especially for class 2 traffic, should be considered.

Further studies at the cell level will investigate intelligent schedulers that adapt better to the traffic mix in order to improve delay sensitive traffic.

ACKNOWLEDGEMENT

The authors gratefully acknowledge the support and financial assistance provided by the UK EPSRC, under the CAMA project GR/L07406.

REFERENCES

[1] ATM-Forum Traffic Management 4.0
[2] ATM Forum Wireless ATM Working Group, BTD-WATM-01.03
[3] M. Fontaine, 1996 "Bandwidth allocation and CAC in ATM networks" Electronics & Comm. Eng. J., Aug. 1996.
[4] Proceedings of the Wireless Broadband Communications Workshop, Brussels, Sept. 1997, AWACS project
[5] BONES Alta Group, Simulation Tool, Designer3.6 1997
[6] ITU-T Recommendation I.350, I.371
[7] J. Sanchez, R. Martinez "A survey of MAC protocols proposed for Wireless ATM", IEEE Network, Nov/Dec 97, pp.52-62.
[8] M. Ajmone Marsan, A. Bianco et Al. "A CAC Algorithm for ATM Networks Supporting Different QoS Classes", COST 257 project, www-info3.informatik.uni-wuerzburg.de/cost/
[9] ACTS project AC085 Magic WAND, http:// www.tik.ee.ethz.ch/~wand/.
[10] S. Sezer, R. Woods, A. Marshall, "Architectural trends in Commercial ATM switches", 4th Communication Network Symposium, 1997, Manchester, pp.49-52.
[11] S. Sezer, F.E. Garcia-Palacios, A. Marshall, R. Woods "Buffer Architectures for Predictable Quality of Service at the ATM layer", Globecom'98, Sydney, Nov. 1998.
[12] D.P. Heyman, A. Tabatabai et al, 1992 "Statistical Analysis and Simulation Study of Video Teleconference Traffic in ATM Networks", IEEE Trans on circuits and systems for video technology, vol.2, No.1, pp.49-59
[13] F.E. Garcia-Palacios, A. Marshall, S. Sezer, D. Chieng, "QoS Analysis of a Wireless ATM Network Access Point", ICC'99, Vancouver, June 1999.
[14] M. Naghshineh, M. Schwartz, "Distributed call admission control in mobile/wireless networks", IEEE JSAC, vol 14, no.4, pp.711-717, May 1996.
[15] William Stallings, "Viewpoint:Self-similarity upset data traffic", IEEE Spectrum, Jan 1997.

PART TWO

Quality of Service in Internet Networks

Chapter 4

Automating Internet Routing Behavior Analysis Using Public WWW Traceroute Services

Selim Gurun and Boleslaw K. Szymanski
Department of Computer Science, RPI, Troy, NY 12180-3590

Key words: Internet, route tracing, routing anomalies, path stability, path symmetry.

Abstract: Growing dependence of commerce, industry and academia on the Internet underlines importance of analyzing its behavior. Changing Internet structure makes the gathered data quickly obsolete, yet the results of analyses are useful in validating network management software and network protocols, as well as in planning extensions and enhancement. In this paper we describe tools for data analysis and an experiment conducted during the summer of 1999 using traceroute for analyzing Internet routing anomalies. We compare our results with Paxton's analysis of 1995's Internet.

1. INTRODUCTION

The Internet performance measurements are useful in validating network management software or protocols. They can also assist in planning new facilities and enhancing existing ones. Finally, such measurements can be used to gather statistics about the traffic, packets and routers. One of the very first studies of this kind investigated the dynamics of routing in the NSFNET backbone network [Ch93]. Data collected during a 12-hour period on August 1992 from all NSFNET nodes on the T1 backbone were used to analyze the frequency of routing table updates. Another study analyzed network inter-domain routing information [LMJ97]. The authors collected BGP routing messages generated by border routers located at five public exchange points in the U.S. over the nine months. An important analysis of routing pathologies in the Internet was presented in [P97]. The author made end-to-end measurements using *traceroute* program between nearly 30 sites and used the results to study routing pathologies, path stability and routing symmetry in the Internet. The author also analyzed Internet packet dynamics by initiating and tracing TCP transmissions between the investigated sites. Data were collected during a few week periods in 1994 and 1995.

In this paper we describe tools for data analysis and an experiment conducted during the summer of 1999 during which *traceroute* measurements were collected and used to analyze the Internet routing anomalies. We closely followed methodology of analysis used in [P97] to easily compare routing problems of 1995's Internet with the problems existing today.

2. DESIGN AND IMPLEMENTATION OF MEASUREMENTS

2.1 Introduction

traceroute utility, available on many platforms, traces the hops on a route from the host to the destination using Time-To-Live (TTL) field of transmitted packets. It relies on the IP requirement that a packet with TTL field smaller than 2 is discarded by all routers except the destination. If a router discards a packet, it also sends an ICMP reply to the host that sent this packet. *traceroute* repeatedly sends packets to the traced destination with increasing values in TTL field, starting with 1. From the ICMP replies that it receives, *traceroute* reconstructs the route followed by the packets. The process stops when the packets sent in one step reach the destination and therefore do not trigger the ICMP reply. *traceroute* also calculates round trip time, and collects errors such as unreachable host, no replies, or unreachable network. *traceroute* output shows routing loops, and other routing pathologies, like fluttering paths. It is also possible to evaluate path symmetry or the average lifetime of paths using this information. The detailed description of *traceroute* utility is given in [JA89] and [P97].

Traditionally, tracing a route from a host machine to a destination requires telnet access to the host machine. Currently, some web sites have installed a CGI interface, which enables anybody to trace routes from these sites to any other host using WWW. A list of traceroute servers accessible via WWW is given in [T1] [T2] and [T3]. The price for using the existing and easily accessible servers was limited control over parameters of the call. Most often, WWW traceroute servers allow only passing the name or IP address of a remote site. Some of the sites allow measurements only between themselves and the host that has directly connected to their web server making them useless to our analysis. Most of the sites use a script that restricts the measurement time to 45 seconds. The adverse effect of this restriction is discussed in the sections analyzing the results in detail. The results of measurements are displayed as HTML pages that are inconvenient for analysis, so we provided a parsing utility.

2.2 Basic Design of Measurements

We created a small model of the Internet, which is formed from hosts at many locations around the world. The selected hosts provide WWW traceroute service that we used to measure the paths to other selected hosts at random times for a long period of time. To make this model represent the Internet well, we selected the sites from geographically diverse and heavily wired locations. Thanks to this selection, even a small number of hosts can span a large number of paths, routers and autonomous systems because the number of paths grows with the square of the number of sites. We refer to the selected sites as *traceroute servers*. The site that initiates *traceroute* is referred to as a *source site* and the destination of this *traceroute* is called a *destination site*.

We wrote a few utilities to facilitate the measurements. *ndc* is a utility that contacts a traceroute server and tells it to measure a route to a destination site and then collects the results of measurements and save them for later analysis. *ndc* is installed on a single workstation in our computer laboratory, creating a centralized design with a single point of failure. In case of a network failure between our site and the world, *ndc* cannot contact traceroute servers. However, the same centralized design was used in [P97], making the results of both studies comparable. The collected results are preprocessed using *preproc* utility to remove measurements having errors and problems discussed later. They are analyzed using *trazer* utility

2.3 Data collecting (ndc) and analysis (trazer) utilities

ndc is a utility that we wrote to automate process of assigning measurements to traceroute servers. It selects a pair of servers, then contacts each of them to request measurements of routes between them. It also receives, parses and saves the results returned by the servers. There is no single way to communicate with a WWW traceroute server. Some servers use ISINDEX field of HTML page, some others use forms and name fields in a different way. Some servers use POST while others use GET method to communicate with their CGI interface. The results are embedded into HTML pages with a design differing from site to site. These differences forced us to make *ndc* re-configurable for each single site.

The utility was designed to be immune to various network problems, because it must work continuously for quite long periods, a few weeks at a time. The main process creates child processes and tells them where they suppose to make measurements. A child makes a connection with the assigned traceroute servers, and when problems are detected, it cleans its leftovers, writes an error log and then exits.

ndc can make measurements in two different time scales, one with the large average inter-measurement time between two specific sites and the other with this time set to the much smaller value. The reason for having two different scales is to keep the size of data collected reasonable without jeopardizing the quality of analysis of results. A short interval between measurements is useful in analyzing route stability and duration of routing errors. However, the number of traced paths is proportional to the square of the number of participating sites, so a short inter-measurement time yields huge amounts of data and introduces significant load to the network.

After preprocessing is complete, the data is analyzed, using a utility named *trazer,* for path stability, routing pathologies and path symmetry. *traceroute* sends three packets at each hop. Sometimes these packets traverse different paths, causing different routers to send a reply. This is not an error or pathology but often a sign of load balancing. However, this also causes a problem during analysis of routing symmetry and stability. One option of *trazer* selects and removes the traceroutes with *ambiguous* paths. Another enables the user to replace router addresses with substitutes, which aids in locating routing errors.

3. ANALYZING RAW DATA AND ERRORS

3.1 Measurements having no hop information

We collected routing data during a two weeks from June 1 to June 14, 1999. We have used 29 sites, which are located in North American, European and Asian-Pacific countries. These sites and their locations are listed in [GS99]. All these sites, except `us4`, continued giving traceroute service until the end of our experiment. Starting on Jun 7-8 midnight, site us4 canceled traceroute service. During the experiment, a total of 13172 traceroutes were attempted. Among these, 754 measurements failed by returning no hops data. The rest of measurements, a total of 12418, were collected and saved successfully. We analyzed 754 failures further using our measurements of path symmetry made for each traceroute.

23 traceroutes were lost because of errors in concurrent writing of the results by *ndc* children to a single file. Fortunately, only 0.17% of all traceroutes were affected by this problem.

104 traceroutes were lost because "connection timed out. Nearly a third of these failures were from three different time intervals and related with three different sites. In the longest of these periods, on June 3, from early morning to noon, a series of 23 connection attempts to us5 had ended in connection time out error. Surprisingly, during this period all the traceroutes conducted from the other sites to us5 have ended successfully! This makes us to believe that an Internet connectivity failure between our site and us5 lasting several hours had caused this problem. Another such period was on June 11, lasting for several hours and causing a series of 7 connection time out failures from tr1. Also the symmetric traceroutes to tr1 failed just on the border of metu.edu.tr domain during this period. However, except the failures discussed above and a series of 7 failures with us7 on June 13, all the other failures were distributed fairly evenly among the sites and persisted for much shorter time periods.

There were 6 failures due to "connection refused by the remote host" error. 45 connection attempts failed with "no route to host" error, out of which 23 were from ca1 and 20 were from de3. Our data shows that from June 6 to June 11, 20 connection attempts to ca1 returned this error. Our observation is that the sites/domains go into an unreachable period, which lasts for a few hours at a time and then recover. One interesting observation is that the connectivity problem generally appears again in less than 24 hours later. As an example, ca1 was unreachable from 21:45 to 23:55 on June 6, then it was unreachable from 13:00 to 18:00 on June 7, and this was repeated on June 8. We discuss this phenomenon again from different perspective in the routing error section.

Table 1. Distribution of Failures

Unsaved/lost	23	Concurrent writing in *ndc* utility
Connection timed out	110	Mostly 3 sites in specific intervals
No route to host	45	Mostly from ca1 and de3
Traceroute service stopped	343	Site stopped supporting *traceroute*
Unknown host	18	IP address of the destination not found
CGI script restrictions/other	215	45 second limitation in most scripts
Total	754	Measurements contained no data

As mentioned in the introduction part of this section, us4 has stopped *traceroute* service on June 7. Because of this stoppage a total of 254 traces had failed. There was also a temporary service outage with us2 that started on June 5 and ended on June 8. During this time, 89 traceroutes failed.

18 traceroutes failed with an error of "unknown host". This was most probably caused by the DNS service problems.

The remaining 215 failures were mostly caused by the 45 second time limit in the CGI script.

3.2 Interrupted Measurements

The limit on the measurement time causes the trace to stop. Such stopped measurements can be identified by the name of the last hop being different from destination. This group of errors contains also traces of routes that are more than 30 hops long, because this is the length limit of *traceroute* tracing capabilities. Hop data of interrupted measurements is useless in analysis of path symmetry and lifetime, but can be used in analyzing routing loops.

Only 175 traces or 1.3% of the total number, suffered from this problem (see table 3). The effect of the time limit is small, except of site tr1. From a total of 454 traceroutes attempted from this site, 152 failed or returned no data and 74 were interrupted because of timing limit, which for this site was 345 seconds, much higher than for the most of other sites. This indicates a serious congestion and delay at this site.

Some hosts are located behind firewalls, which drop 'undefined packages'. These hosts are seen as unreachable to *traceroute* utility. For example, among the 422 measurements directed to se1, 417 ended in *ITcenter-gw.DMZ.Umea.SE* and the remaining 5 were interrupted because of the time limit. Since the circular references after firewall node do not indicate routing loops we excluded them from the routing loop analysis.

Table 2. Distribution of Failures (Interrupted Measurement) by site

Site	*#*	*% of total*	*Site*	*#*	*%of total*	*Site*	*#*	*%of total*
us4	256 (1)	33.95	us7	9 (4)	1.19	au1	3 (2)	0.40
tr1	152(74)	20.16	hk1	9 (3)	1.19	at1	3 (8)	0.40
us2	91 (3)	12.07	se1	7 (5)	0.93	us9	2 (2)	0.27
de1	67 (1)	8.89	su1	5 (3)	0.66	us3	2 (3)	0.27
ca1	31 (2)	4.11	sp1	4(10)	0.53	us1	2 (2)	0.27
de3	25 (2)	3.32	nz1	4 (1)	0.53	ch1	2 (2)	0.27
us5	24 (3)	3.18	us12	3 (5)	0.40	us8	1 (5)	0.13
it1	11 (6)	1.46	jp1	3 (8)	0.40	us13	1 (1)	0.13
us6	10 (7)	1.33	cz1	3 (2)	0.40	us10	1 (3)	0.13

Table 3. Trace statistics (* denotes exclusion of 23 unsaved attempts, ta=traces attempted, f=traces failed, i=traces interrupted and %e=percentage of incomplete traces)

Site	ta*	f*	i	%e	Site	ta*	f*	i	%e	Site	ta*	f*	i	%e
at1	466	3	8	2.4	jp1	437	3	8	2.5	us5	477	24	3	5.7
au1	448	3	2	1.1	nz1	496	4	1	1.0	us6	442	10	7	3.8
ca1	402	31	2	8.2	se1	460	7	5	2.6	us7	481	9	4	2.7
ch1	538	2	2	0.7	sp1	459	4	10	3.1	us8	485	1	5	1.2
cz1	422	3	2	1.2	su1	419	5	3	1.9	us9	499	2	2	0.8
de1	459	67	1	14.8	tr1	454	152	74	49.8	us10	412	1	3	1.0
de2	423	0	5	1.2	us1	448	2	2	0.9	us11	468	0	2	0.4
de3	378	25	2	7.1	us2	404	91	3	23.3	us12	445	3	5	1.8
hk1	468	9	3	2.6	us3	481	2	3	1.0	us13	433	1	1	0.5
it1	446	11	6	3.8	us4	499	256	1	51.5	total	13149	731	175	6.9

4. ROUTING PATHOLOGIES

4.1 Permanent routing loops

Assume that we are tracing a path from a site A to a site B and the routers in the path are named R_1 to R_n. We assume that there is a **persistent** routing loop if the reported path is $R_1, R_2, \ldots R_i, R_{i+1}, \ldots R_k, R_i, R_{i+1}, \ldots R_k, \ldots$ and the trace ends before reaching site B because of the 30 hops limit. In other words, a **persistent** routing loop is a loop that does not resolve during a *traceroute* time. A loop in *traceroute* does not necessarily mean that there is a real forwarding loop. In [P97], a loop is classified as a real forwarding loop if the cycle is repeated three times. We have used the same rule in our analysis.

There were 12 persistent routing loops. In all twelve cases, a routing loop in a traceroute from site A to B corresponded to failures of measurements from site B to A with error messages, like "no route to host" and "connection timed out". Most of the loops were directed to site ca1. This site became unreachable between June 3 and 11. Three of the permanent loops affected only one measurement. The measurements just before and after these have no loops. One of the loops, from su1 to jp1, has lived for two traces measured 10 minutes apart. None of the loops spanned two autonomous systems.

A temporary routing loop is the one that is resolved before *traceroute* ends. It may cause congestion by producing duplicate packets in the system [P97]. We found 11 temporary routing loops in our test data, 7 of them were either started from or directed to tr1.

4.2 Fluttering Paths

Fluttering refers to rapidly varying routing [P97]. More than one router name for a hop indicates that the path has changed during the trace time, which usually lasts tens of seconds. Fluttering may be a side effect of load balancing. However, it might introduce problems described in [P97] and [V97].

In our study we found that fluttering is very frequent in the Internet. Approximately 25% (3011 traceroutes) of all the traces contains at least one hop with two different routers. However, the distribution of fluttering paths is not equal. Nearly half (1501 traceroutes) of the fluttering paths either have jp1 or se1 as source or destination site. We also found that most of the affected paths have only one or two hops involved in fluttering, a case which we call a *minor fluttering*. To understand the distribution of fluttering, we assigned weights to fluttering paths. If a hop had more than one router reported, we assigned 1 to it, otherwise it was assigned the weight 0. The weight of a traceroute is the sum of weights of all hops. The distribution below shows that only a small percentage of paths suffer from *heavy fluttering*.

Table 4. Distribution of fluttering paths

Weight	=1	=2	=3	=4	>4	Total
%fluttering	2208 73%	607 20%	119 3.9%	44 1.5%	33 1%	3011 100%
%all traces	17%	4.8%	0.95%	0.35%	0.26%	24%

The negative effects of fluttering increase if the paths are significantly different from each other. Hence, we studied fluttering in Autonomous System (AS) granularity. This means that we assigned the weight 0 to a fluttering hop if the fluttering routers were in the same AS. Replacing router numbers with AS numbers in traceroute weight computation yielded the results shown in Table 5.

Table.5. Distribution of fluttering paths in AS granularity

Weight	=1	=2	=3	=4	>4	Total
%fluttering	448 78%	95 16%	25 4.3%	2 0.3%	3 0.4%	573 100%
%all traces	3.6%	0.76%	0.2%	0.01%	0.02%	4.6%

4.3 Unreachable messages

173 traceroutes returned message "*access administratively prohibited*". All of these traceroutes were directed to site se1. After removing failed and interrupted measurements, 61 traceroutes returned either "host unreachable" or "network unreachable" message with the following distribution among sites: ca1 24, de3 19, su1 4, us7 3 and others 11.

5. ROUTING STABILITY

Following [P97], we analyzed routing pathology using two basic notions: *prevalence* and *persistence* of routing. The first one defines the probability that a route *r* observed at time *t* is observed again at time t+s. In other words prevalence defines how likely a route is to be used again. The persistence defines the average lifetime of a path and it is not easy to measure based on information collected over small time periods. We evaluated stability in two different granularities to distinguish between minor and major path changes. Using bare IP names, we conducted *host granularity* analysis and then replacing IP names with AS numbers we made *AS granularity* analysis. A path rapidly switching between autonomous systems indicates much more serious problems then a path switching within the same AS.

Before analyzing path stability we removed incomplete traceroutes missing one or more hop data (eliminating 660 traceroutes) and ambiguous traceroutes with more than one router for any hop (removing another 2862 traceroutes). That left 8896 traceroutes for host granularity analysis. In AS granularity the number of ambiguous traceroutes decreases, leaving 11226 traceroutes for the corresponding analysis.

5.1 Prevalence

Our study includes measurements of 812 potential paths, i.e., *source-destination pairs*. After eliminating irrelevant and ambiguous traceroutes we had data on 727 paths. For each path, we calculated the ratio of number of times the dominant route was used to the total number of time the path was traversed. For example, assume that the path from A to B was traversed 9 times along the following routes:

R_1, R_2, R_2, R_1, R_3, R_1, R_1, R_1 R_1.

Then the prevalent route is R_1 and the prevalence of R_1 is 6/9 or 0.66.

The result of our analysis shows that 421 of 727 (58%) paths used the prevalent route at least 50% of time.

When we repeat the analysis at the AS granularity, results change dramatically. From 787 analyzed paths, only 3% (30 paths) had used the prevalent route less than 50% of time.

5.2 Persistence

We analyzed persistence in five different time scales: 60 sec., 10 min., 1 hour, 10 hours and more than 10 hours. For each category, we selected routes that changed in this time, and then calculated the ratio of varying paths to stable ones. We also studied which source/destination pairs are likely to switch frequently and which pairs are more long-lived. We again used two different granularities, host and AS granularity, as we did for prevalence. We used the same data that we used for prevalence. The results, shown in Table 6, are cumulative, i.e., 10-min. scale also covers the data for 60-sec. scale. The number of route changes counts number of changes from one router to another in a route. About a third of route changes at any scale are inter-AS route changes.

At 60 sec. scale all changes involved (at1,us11) and (au1,us5) pairs with low use of prevalent paths. Out of 60 changes for 10-min. scale, 44 routes started at at1 or de2 and 20 ended at us11. Nearly half of the route changes at 1 hour scale started at five sites (at1 14%, ch1 11%, us5 10%, de2 8% and au1 8% with a total of 54% of all changed routes), that were involved in a few routes, observed. The same sites were responsible for 23% of all route changes in AS granularity.

Table 6. Persistence in Host (AS) granularity

Scale	#of Routes	#of Route Changes
<=60 sec	47 (68)	2 (0)
<=10 min	405 (560)	60 (9)
<=1 hour	1900 (2551)	475 (117)
<=10 hours	5717 (8180)	1879 (525)
10 hours+	8897 (12013)	3311 (1023)

6. ROUTING SYMMETRY

Following [P97], we define two routes between a pair of sites as symmetric if they have the same number of hops, n, and $\forall i, 1 \le i \le n : r_i = r_{n-i}$, where r_i denotes a router.

We studied routing symmetry only in AS granularity, using only unambiguous traceroutes (i.e., those without errors or missing hops, with single router traversed by each hop and uninterrupted). This left us with 6624 traceroutes or 3312 traceroute pairs.

Table 7 with the results of analysis shows that only 36% of all measurements were symmetric in AS granularity. If we include the paths having only one hop difference, this ratio increases to 42.8%. In addition, 47% of pairs have different number of hops and are not included in this count. Another conclusion is that generally each pair maintains a fairly constant level of asymmetry. For example, we have 15 observations with asymmetry level equal to 7 and all of them included de1 either as source or as destination. To support the above conclusion, for each level of asymmetry we computed percentage of asymmetry associated with 50% of pairs that have most asymmetric paths. This value is about 80% at all levels of asymmetry, indicating that path asymmetry is quite unevenly distributed among hosts. Hence, paths are stable and asymmetric and each pair produces pretty constant level of asymmetry.

Table 7. Classification of pairs in terms of symmetry of their paths

pairs with different #of hops	1561	47%
asymmetric pairs with equal #of hops	557	17%
symmetric pairs	1194	36%
total number of pairs	3312	100%

Table 8. Details of pairs having equal number of hops (w=weight of asymmetry, N=number of measurement, and P=percentage of total measurements)

W	N	P	Comments
w=0	1194	36	Symmetric pairs
w=1	424	12.8	
w=2	83	2.5	
w=3	24	0.72	w>=3 is less than %1
w=4	3	0.09	
w=5	8	0.24	
w=6	0	0	
w=7	15	0.45	site del in all w=7
total	1751	52.86	

Table 9. Details of asymmetric pairs

difference between #of hops	#of pairs	% of all pairs
h=1	566	17
h=2	560	16
h=3	250	7
h>3	185	5.5

7. SUMMARY OF RESULTS

During a two weeks period, we collected 13172 traceroutes from 29 sites. These 29 sites introduced 812 possible paths. After eliminating the erroneous 754 traceroutes, we were left with 12418 measurements to analyze routing pathology, stability and symmetry.

We found 12 persistent and 11 temporary routing loops. However, our results may be well underestimating the real number of errors, because we have lost 390 measurements (215 with no data and 175 measurement interrupted in the middle) as a result of a time limit that restricts the *traceroute* measurement time to 45 seconds. Unfortunately, in case of an Internet connectivity problem, a *traceroute* is likely to take more time, which means some of interesting routing problems are lost with these 390 measurements.

We studied fluttering and path stability in two different granularities. First, we made our analysis based on each host visited and then on each autonomous system visited. We called the former *host granularity* and the latter *AS granularity.* Generally, a problem affecting more than one AS is much more serious then a problem inside a single AS.

We found that the ratio of fluttering paths in host granularity is 24% and it drops in AS granularity to 4.6%. However, not all the paths suffer evenly from fluttering. Some sites are much more likely to introduce varying paths. Nearly half of the fluttering paths had jp1 or se1 either as a source or as a destination.

We studied path stability in two different aspects, prevalence and persistence. The most frequently used route is called a prevalent route and the probability of using prevalent path is referred to as prevalence. The persistence measures the average lifetime of paths. The results of prevalence analysis in host granularity show that for half of the paths, the probability of using the prevalent path is 58%. In AS granularity this ratio increases to 97%.

Calculating the average lifetime of paths seems a much more difficult problem. We separated the data into different scales and then studied each group. The ratio of route changes to number of routes is an increasing function. We found that in 60 seconds scale there were 2 route changes among the 47 routes observed. In one hour scale, there were 475 route changes among 1900 routes and in 10 hour scale, there were 1879 route changes among 5700 routes.

We studied routing symmetry only in AS granularity level, because in host granularity almost all routes are highly asymmetric. We found 36% of all routes are symmetric. 12% of routes differ only in one AS and 5% of them differ in more than one AS. For 47% of traceroute pairs each traceroute had different number of hops so we did not calculate the asymmetry weight for them. We also noticed an interesting asymmetric path between two sites located in Germany and Italy in which packets from Germany go to USA before reaching Italy!

Our motivation for this study was to evaluate the Internet in terms of routing behavior and to compare our results to the previous studies [P97].

ACKNOWLEDGEMENT

This work was supported in part by DARPA grant F19628-98-C-0057. The content of this entry does not necessarily reflect the position or policy of the U.S. Government---no official endorsement should be inferred or implied.

REFERENCES

[CH93] B. Chinoy, "Dynamics of Internet Routing Information," Proceedings of SIGCOMM'93, pp. 45-52, September,1993.

[GS99] S. Gurun, B. Szymanski "Internet Routing Behavior Analysis Using Traceroute," Technical Report, Dept. Computer Science, Rensselaer Polytechnic Institute, Troy, NY, TR99-11, http://www.cs.rpi.edu/~szymansk

[JA89] V. Jacobson, traceroute, ftp://ftp.ee.lbl.gov, 1989.

[LMJ97] C. Labowitz, G.Malan, F. Jahanian, " Internet Routing Instability," Proceedings of SIGCOMM '97, September 1997.

[P97] Vern Paxson, "Measurements and Analysis of End-to-End Routing Internet Dynamics," Ph.D. dissertation, University of California, Berkeley, April 1997.

[T1] http://www.traceroute.org

[T2] http://www.netaxs.com/~rmsolino/tech/tracert.htm

[T3] http://www.geocities.com/Athens/4273/traceroute.html

[V97] Vern Paxson, "End-to-end routing behavior in Internet," IEEE/ACM transactions on Networking 5, page 601-615.

Chapter 5

An Evaluation of Different Marking Algorithms in Assured Forwarding Service

T. Yang*, Z. Chen*, A. Hafid&, D. Makrakis*

**School of Information Technology & Engineering, University of Ottawa*
Colonnel By Hall (A617), P.O.Box 450 Stn A, Ottawa, Ont., K1N 6N5, Canada
{tyang,zhchen,dimitris}@site.uottawa.ca
&University of Western Ontario, Computer Sciences Dept., London, Ontario, Canada

Key words: Assured Forwarding, Differentiated Service, TSW, TCM, Improved-TSW.

Abstract: In this paper, we overview two popular marking algorithms that are in use in networks that support the Assured Forwarding (AF) Service of Differentiated Services (DS); the algorithms are: Time Sliding Window (TSW) algorithm and Three Color Maker (TCM) algorithm. We present the simulations we have performed to evaluate and compare the performance of these two algorithms in an AF service-capable network. Then, we propose a new marking algorithm, called Improved-TSW algorithm, which is designed by combining TSW algorithm with a token bucket. A performance evaluation, using simulations, show that the new algorithm provides a better performance, in an AF service-capable network, than the other algorithms.

1. INTRODUCTION

There are many algorithms available for metering and marking packets at the boundary nodes of a differentiated service (DS) [1] domain, among which, Time Sliding Window algorithm [2] and Three Color Marker algorithm [3] are the two most popular ones. The objective of this paper is to evaluate and compare the performance of these two marking algorithms in an AF service-capable network [4] and propose a new marking algorithm

& A. Hafid is currently with Telcordia Technologies (formerly Bellcore), NVC3X359, 331 Newman Springs Road, Red Bank, NJ, 07701; e-mail: ahafid@research.telcordia.com

named Improved-TSW algorithm. We performed a series of simulations to evaluate the marking algorithms using OPNET 5.1 [5], a network simulation tool developed by Mil3. In sections 1.1 and s 1.2, a brief description of two existing marking algorithms (TSW and TCM) is presented.

1.1 Time Sliding Window Algorithm

Time Sliding Window (TSW) [2] algorithm is a rate estimator algorithm. TSW provides a smooth estimate of the TCP sending rate over a period of time. With the estimated rate, Avg_rate, the marking algorithm can mark packets as OUT (out of profile) packets once the traffic exceeds a certain target rate.

A rate estimator is used to smooth out the bursts of TCP traffic as well as to be sensitive to instantaneous sending rate. The design of TSW is very simple. TSW maintains three state variables: Win_length, which is measured in units of time, Avg_rate, the rate estimate upon each packet arrival, and T_front, which is the time of last packet arrival. TSW is used to estimate the rate upon each packet arrival, so variables Avg_rate and T_front are updated each time a packet arrives, but Win_length is pre-configured when the profile meter is installed (in our simulations, we chose the Win_length on the order of the RTT of the TCP connections according to [2]). Follows is a description of the main steps of TSW algorithm.

```
Initialization:
   Win_length = a constant;
   Avg_rate = connection's target rate, R_T;
   T_front = 0;
For each packet with size pkt_size arrival at time arrive_time:
         If it is an IN (in profile) packet
               Bytes_in_TSW = Avg_rate * Win_length;
               New_bytes = Bytes_in_TSW + pkt_size;
               Avg_rate = New_bytes / (arrive_time - T_front + Win_length);
               T_front = arrive_time;
               If Avg_rate > 1.33 * R_T
                     Remark this packet as OUT packet;
               Endif
         Endif
```

1.2 Three Color Marker Algorithm

The Three Color Marker (TCM) [3] algorithm is a kind of token bucket algorithm. It meters an IP packet stream and marks its packets either green, yellow, or red. Marking is based on a Committed Information Rate (CIR) and two associated burst sizes, a Committed Burst Size (CBS) and an Excess Burst Size (EBS). CIR is measured in bytes per second for IP packets. The CBS and EBS are measured in bytes. A packet will be marked as green if it doesn't exceed the CBS, yellow if it does exceed the CBS, but not the EBS, and red otherwise.

The Meter meters each packet and passes the packet and the metering results to the Marker; in our simulations, the Meter operates in Color-Aware mode [3]; this means that the Meter assumes that the packets have been pre-marked. Then, the Marker remarks, if necessary, the packet according to the results of the Meter.

The Meter measures the packet stream by using two token buckets, C and E, that both share the common rate CIR. The maximum size of the token bucket C is CBS and the maximum size of the token bucket E is EBS. The token bucket C and E are initially (at time 0) full, i.e., the token count TC(0) = CBS and the token count TE(0) = EBS. Thereafter, the token counts, TC and TE, are updated CIR times per second as follows:

```
If TC < CBS
    TC = TC +1;
Else if TE < EBS
    TE = TE +1;
Else
    Neither TC nor TE will be incremented;
```

When a packet of size B bytes arrives at time t, the following operations are performed:

```
If the packet has been pre-marked as green and TC(t) – B >=0
    The packet is marked green here and TC = TC – B;
Else if the packet has been pre-marked as green or yellow and TE(t) – B >= 0
    The packet is marked yellow here and TE = TE – B;
Else
    The packet is marked red here and neither TC nor TE will be changed;
```

The Marker reflects the metering results by setting the DS field of the packet to a particular codepoint [6]. In case of the AF service, the color can be coded as the drop precedence of the packet. In the simulations we performed, we used only two drop precedences (IN and OUT), so we slightly changed the TCM algorithm described above to fit our simulations. We used only one token bucket; this means that marking is based on CIR and CBS only: a packet is marked IN if it doesn't exceed the CBS, OUT if it exceeds the CBS.

The rest of the paper is organized as follows. Section 2 presents the network architectures and configurations used for the simulations; Section 3 gives the evaluation and comparison of TSW and TCM marking algorithms. Section 4 presents the new marking algorithm, called Improved-TSW marking algorithm; comparison of the performance of the three algorithms is also presented. Section 5 concludes the paper.

2. NETWORK ARCHITECTURE AND CONFIGURATIONS

2.1 Network Architecture

The network architecture shown in the Figure 1 is used. Traffic is sent from the clients to the servers. Client 1, Client 2, Client 3, and Client 4 are FTP clients while Client 5 is an email client. Server 1, Server 2, Server 3, and Server 4 are FTP servers while Server 5 is an email server. The reason behind this choice is that the FTP traffic is the main part of the total traffic in this simulation and the Email traffic is just the background traffic. Traffic is sent from Client i to the Server i (1<=i<=5). FTP traffic uses TCP as a transport protocol and all the FTP packets are marked with IN (in profile) by the FTP clients; packets marked with IN have low drop precedence and are protected by the AF service-capable network from other best-effort packets. Email traffic uses UDP as a transport protocol and all the Email packets are marked with OUT (out of profile) by the Email client; packets marked with OUT have high drop precedence and are dropped with high probability when there is congestion in the network. All the FTP clients have the same configurations. Marker (see Figure 1) is the boundary node of this AF service-capable network; marking algorithms are performed at Marker. In our simulations, we used TSW, TCM (see Section 1) and Improved-TSW (see Section 4) algorithms to meter and remark, if necessary, FTP traffic against the profile for this traffic. We assume that all the acknowledgements are sent from servers to clients without losses.

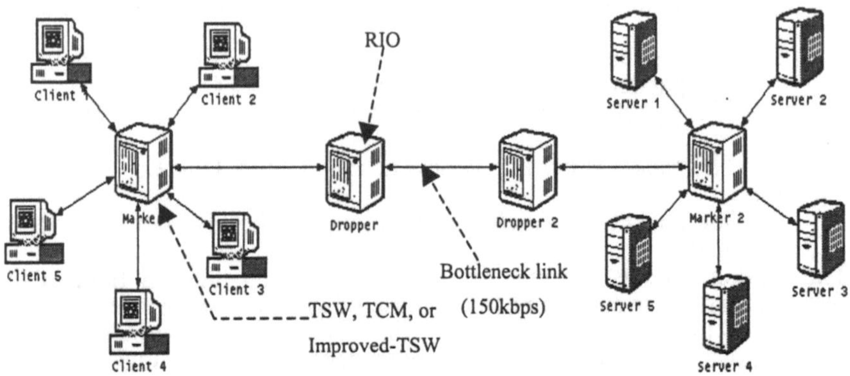

Figure 1. DS Domain Network Architecture for Simulations

2.2 Configurations for Simulations

2.2.1 Basic Simulation Configurations

The followings are parameters that are used in all simulations presented in this paper. Simulations are run for 1000 seconds in OPNET; it is the time needed for the network to be stable. Values for IP packet and TCP segment size are the default values set in OPNET. Application segment size is the maximum segment size the application layer can send to the lower layer. Application segment size should not be too big; otherwise, it will take more time for senders to fragment it and if one fragment in this segment is lost, receivers will not receive the whole application segment. Application segment size should not be too small either; otherwise, the system efficiency will suffer (i.e., unacceptable protocol overhead). We selected a value of 8000 bytes for the application segment size. Application start time is the time the application starts sending packets; we use the OPNET default value that is equal to 100 second. This means the FTP/e-mail traffic is sent out after 100 seconds simulation time.

Simulation Duration: 1000 sec IP Packet Size: 576 bytes
TCP Segment Size: 536 bytes Application Segment Size: 8000 bytes

Application Start Time: 100 sec

2.2.2 FTP Client Attributes

Every FTP client generates traffic at a rate equal to 30Kbps at the application layer. This is the rate the ISP (Internet Service Provider) will try to satisfy for every client. Let us note that the actual traffic (i.e., traffic in the network) will be a little bit above 30Kbps when we take into account TCP and IP headers. The average file size should not be too small; otherwise, there will be many small FTP acknowledgements and requests in the network.
File Transfer Rate: 900 files/hour (modeled using a Poisson distribution)
Average File Size: 15000 bytes (modeled using a normal distribution)

2.2.3 Email Client Attributes

The Email client generates traffic at a rate equal to 80Kbps at the application layer. Let us note that the actual traffic (i.e., traffic in the network) will be a little bit above 80Kbps when we take into account UDP and IP headers. Email traffic will not have a profile at the boundary node of our DS domain. The total amount of TCP and UDP traffic will be larger than the bandwidth of the bottleneck link of the network. The objective is to evaluate the traffic performance of all these types of traffic under congestion.
Send Rate: 1800 messages/hour (modeled using a Poisson distribution)
Average Email Size: 20000 bytes (modeled using a normal distribution)

2.2.4 Links' Characterisation

For sake of simplicity, we assume that all links except the bottleneck link have enough bandwidth to accommodate all traffic generated by client or servers. We set the bandwidth of the bottleneck link (i.e. link between Dropper and Dropper 2 in Figure 1) to 150Kbps; the total traffic will be above 200Kbps. This setting will allow us to analyze the traffic performance of AF service-capable network when the congestion is severe.
Data Rate of all links except the bottleneck link: DS3 (44.736Mbps)
Data Rate of the bottleneck link (the link from Dropper to Dropper 2): 150000bps

2.2.5 Time Sliding Window Algorithm

The target rate is the fraction of FTP traffic the ISP is willing to protect. Actually, FTP clients, altogether, will generate traffic of a rate equal to 120 Kbps at the application layer. TSW algorithm will use this target rate to meter and remark the FTP traffic. When the target rate is exceeded, incoming packets will be remarked as OUT packets.
Target Rate: 120000bps

2.2.6 Three Color Marker Algorithm

The CIR is the token generation rate for the token bucket. The bucket size and the token generation rate for this token bucket are computed based on

the allowed burst size and the target rate of this traffic that are set in the profile for this traffic.
Bucket Size: 64000 bits CIR: 120000bps

2.2.7 Improved Time Sliding Window Algorithm

Improved Time Sliding Window Algorithm is a new marking algorithm that combines TSW algorithm and a token bucket; a detailed description of this algorithm is presented in Section 4. The target rate is set for TSW algorithm and the CIR and Bucket Size are set for the token bucket.
Target Rate: 120000bps Bucket Size: 64000 bits
CIR: 120000bps

2.2.8 RIO (RED [7] with IN and OUT) Algorithm

RIO stands for RED algorithm with IN/OUT bit. RIO uses twin RED algorithms for dropping packets, one for IN packets and one for OUT packets. In RIO [2], packets with different drop priorities will have different thresholds and different probabilities to be dropped. The settings of the parameters of RIO algorithm are aimed to produce simulations where OUT packets have higher dropping probability than IN packets.
Drop Probability for IN Packets (P_{max_in}): 0.03
Drop Probability for OUT Packets (P_{max_out}): 0.08
Maximum Threshold for IN Packets (max_in): 40 packets
Maximum Threshold for OUT Packets (max_out): 30 packets
Minimum Threshold for IN Packets (min_in): 20 packets
Minimum Threshold for OUT Packets (min_out): 10 packets
Queue Weight: 0.002
Queue Length: 60 packets

3. TCM VS. TSW

In Figure 2, the curve named best-effort quantifies the amount of FTP traffic successfully received by the receivers in current best-effort network. The curves named tcm, tsw and improved-tsw quantifies the amount of FTP traffic successfully received by the receivers in AF service-capable network using TCM algorithm, TSW algorithm and improved-TSW (at Marker) respectively.

Figure 2 shows that the amount of FTP traffic, successfully received by the receivers, has increased using AF service-capable network instead of best-effort network (Y-axis represents the amount of FTP traffic that has been successfully transferred; the unit of Y-axis is 10000 * bytes/sec). Based on these observations, we can state that the TCP traffic has been protected in AF service-capable network from the non-adaptive UDP traffic. Next, an evaluation of the performance of different marking algorithms is presented.

In the current simulations, the same packet rate is configured for both marking algorithms, i.e., Three Color Marker (TCM) algorithm and Time Sliding Window (TSW) algorithm. In TCM, CIR (Committed Information Rate) is set to 120 Kbps; in TSW, Target Rate is set to 120Kbps as well. Thus, we are able to analysis the performance of these two algorithms under the same conditions, e.g., if a customer signs up for 120 Kbps with an ISP, we aim to identify the marking algorithm that could provide better service to this customer.

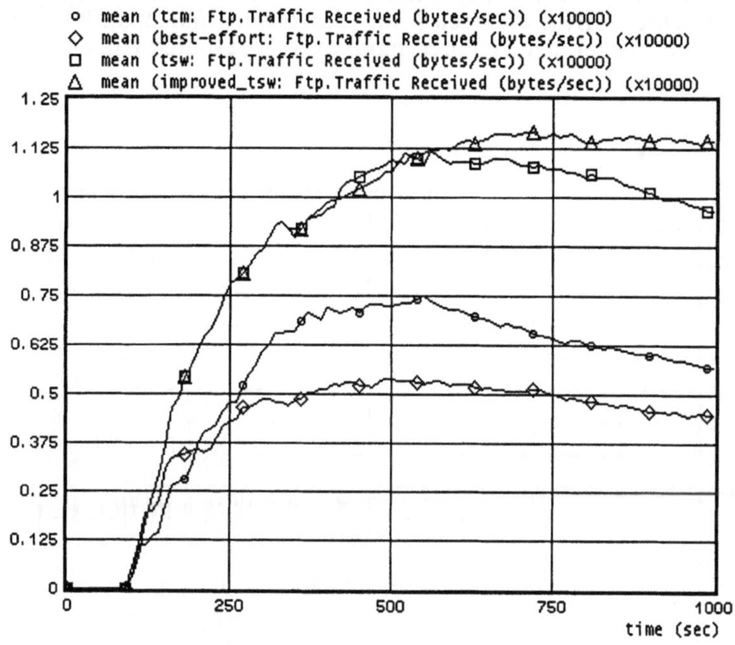

Figure 2. FTP traffic received

Figure 3 shows the number of IN packets remarked to OUT packets at the boundary node Marker using these two algorithms (the unit of Y-axis is packets/sec). In these simulations, we generated very bursty traffic. Although the TCM algorithm allows bursts in the traffic, it tolerates only small bursts. If the traffic burst is very high, i.e., the peak rate of the traffic is much higher than the subscribed rate, the bucket of TCM algorithm will become full very quickly. Then the upcoming IN packets will be remarked as out of profile packets. Figure 3 shows that when using TCM algorithm IN packets change to OUT packets continuously. The reason behind this is that the token generation rate of the bucket in TCM algorithm is only 120Kbps, but the peak rate, i.e., burst, of the traffic is much higher than 120Kbps. Furthermore, the generated traffic includes a large number of bursts making the bucket "always" full when packets arrive. A solution is to increase the bucket size in TCM to allow bigger bursts in the traffic. This is based on the computation of the burst size: Burst Size = CIR * T + Bucket Size, where T is the time for the burst. Nevertheless, since the CIR is generally much lower than the peak rate (burst) of the traffic, the recovery rate of the bucket will be still much lower than the peak rate. Then, after the bucket reaches its full, the upcoming packets will be remarked continuously although the bucket size is bigger than before. Figure 4 shows the mean value of the number of packets that change from IN to OUT at boundary node Marker with bucket sizes of 64K and 160K respectively. After increasing the bucket size from 64K to 160K, fewer packets are remarked to OUT at the Marker. However, the rate of IN to OUT is still high; Rate of IN to OUT = total packets remarked from IN to OUT / total IN packets. Figure 5 illustrates that the traffic performance with the bucket size of 160K is slightly better than the traffic performance with bucket size of 64K. However, increasing the bucket size does not improve the performance as much as increasing CIR (see Figure 6 in which CIR is changed from 120Kbps to 160Kbps). Furthermore, increasing the bucket size requires more system resource (memory) of the router.

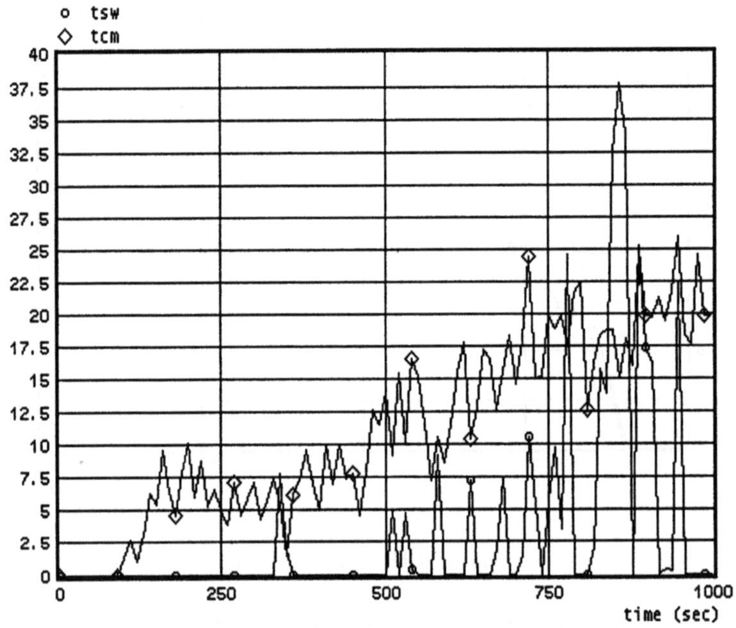

Figure 3. IN packets change to OUT packets at Marker by using TSW and TCM algorithm

Actually, there are two techniques that can be used to improve the traffic performance when using TCM algorithm.

With the first technique we should let the packets pass a shaper before entering the TCM marker. By using a shaper, both the burst size and the peak rate of the traffic are under control. Another one or two token buckets (one token bucket is used in TCM algorithm) combined with this shaper will allow to control the traffic's burst size and peak rate. After passing the shaper, the traffic will not have big bursts; thus, fewer packets will be remarked to OUT when using TCM algorithm. At the mean time, we can let traffic with low burst enter the AF service-capable network to release the burden of interior routers (e.g. less buffer size may be used in interior routers). This will help the traffic behaviour into the AF service-capable network to be more predictable. For example, in our current simulations, even if we mark a burst of IN packets into our AF service-capable network, the IN packet will likely be dropped in the interior nodes because of resource, i.e., buffers, shortage at the Dropper (queues are full). However, if we remove the big bursts from the traffic using a shaper, this problem will disappear. A shaper will be added in front of the marker, to improve the traffic performance of the AF service-capable network, in our future simulations.

With the second technique we should use two token buckets (instead of one) at the boundary nodes to meter and mark traffics. One bucket is used to control the average rate of the coming traffic; there is only one token in this token bucket. The CIR of this token bucket is the average rate of the traffic. When a packet arrives and there is a token in this token bucket, this packet enters the AF service-capable network without passing the second token bucket; otherwise the packet enters the second token bucket. The second token bucket is used to control the bursts of the coming traffic; the bucket size of the second token bucket controls the burst size of the traffic and its CIR controls the frequency of the bursts in that traffic. If a packet enters the second token bucket and there are no enough tokens in this bucket, the packet will be remarked as OUT packet. By using two token buckets to meter and mark traffics at the boundary nodes, we should get better traffic performance.

Figure 3 shows that much fewer IN packets have been remarked to OUT packets by using TSW algorithm than TCM algorithm; in fact, in TSW algorithm only the packets with the average rate above 1.33*R_T (1.33 * 120Kbps) are remarked as OUT packets. The bursts of IN packets whose rate is below 160Kbps pass the Marker without remarking. Figure 3 also shows that all the IN packets that have been changed to OUT packets by using the TSW algorithm are those packets whose average rate is above 160Kbps. This means that only big bursts are remarked to OUT packets in the TSW algorithm.

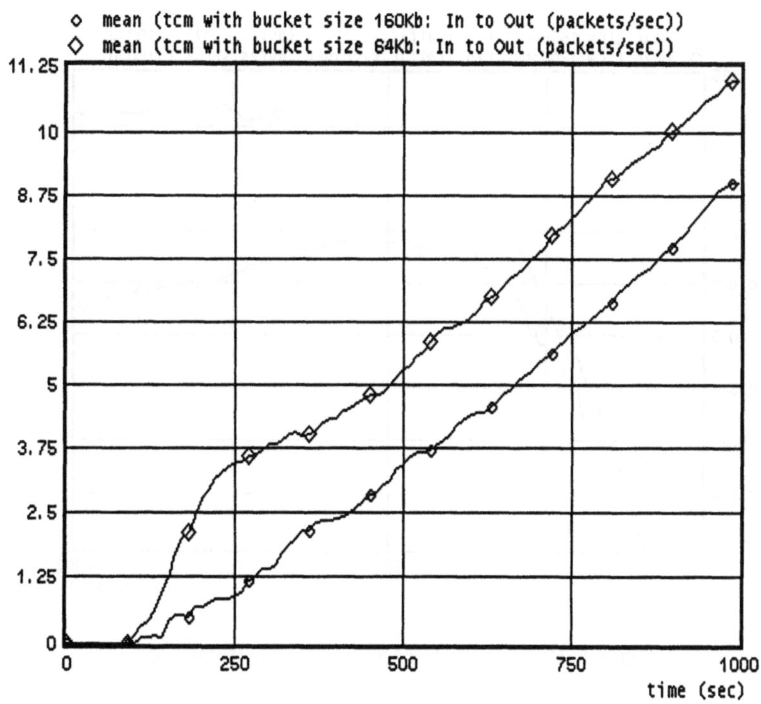

Figure 4. Packets remarked from IN to OUT with different bucket size

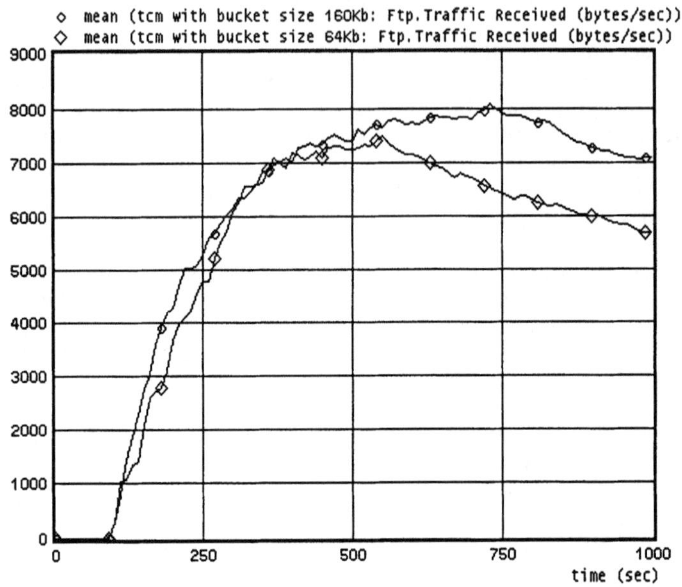

Figure 5. FTP traffic received with different bucket size

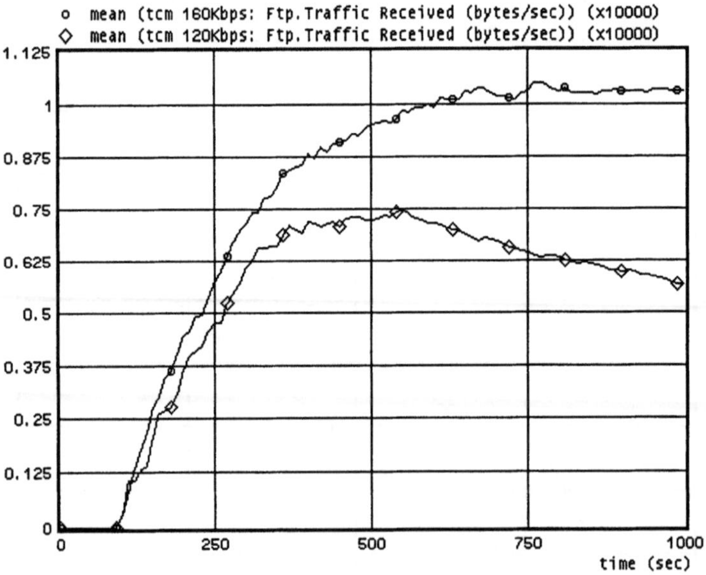

Figure 6. How CIR in TCM affect traffic performance

Figure 2 shows that TSW algorithm performs better than TCM algorithm. This means that more FTP traffic has been successfully received by the receivers when using TSW algorithm. This is rooted in the fact that less IN packets have been remarked at the boundary node Marker in TSW than in TCM algorithm. Therefore, more IN packets (FTP packets) have been allowed into our AF service-capable network and thus protected from OUT packets (Email packets). In fact the simulations show that when the traffic's burst is very high and the congestion is very severe in the bottleneck link, using TSW algorithm results in better traffic performance than using TCM algorithm. However, it is not always true that TSW algorithm performs better. For example, when metering a CBR (Constant Bit Rate [8]) source, a profile meter using TSW algorithm would allow the source to transmit at a sustained rate higher than the contracted one. TCM algorithm does not permit this [9].

4. IMPROVED-TSW ALGORITHM

In order to get better traffic performance than TSW and TCM, we designed a new marking algorithm by combining TSW algorithm and a token bucket. Figure 3 shows that when using TSW algorithm, the packets remarked from IN packets to OUT packets are those ones whose rate is above $1.33*R_T$. A way to improve the performance of this algorithm is to reduce the number of packets remarked from IN packets to OUT packets; token bucket can be used to realize this. If an IN packet's rate is above $1.33*R_T$, we first examine whether there are enough tokens in this token bucket. If there are tokens left in the bucket, the packet will be not remarked; otherwise; the packet will be remarked as OUT packet. Thus, more IN packets will be allowed to enter the AF service-capable network. This new algorithm is a combination of TSW algorithm and a token bucket algorithm: we call it Improved Time Sliding Window algorithm (Improved-TSW).

<u>Improved Time Sliding Window Algorithm</u>

```
Initially:
   Win_length = constant;
   R_T = Target Rate;
   Avg_rate = R_T, Connection's Target Rate;
   T_front = 0;
   CBS = Bucket Size;
   TC = CBS;
   Update time = 1 / CIR;
For each Update time:
         If TC is less than CBS, then
              TC is incremented by one;
         Endif
For each packet with size pkt_size arrival at time arrive_time:
         If it is an IN packet
                Bytes_in_TSW = Avg_rate * Win_length;
                New_bytes = Bytes_in_TSW + pkt_size;
                Avg_rate = New_bytes / (arrive_time - T_front + Win_length);
                T_front = arrive_time;
                If Avg_rate > 1.33 * R_T and TC > pkt_size
                      TC = TC - pkt_size;
                Else
```

If Avg_rate > 1.33 * R_T and TC <= pkt_size
Remark this packet as OUT packet;
Endif
Endif
Endif

TC is the current token number in the token bucket; CBS is the token bucket's size; T_front is the time of the last packet arrival; Avg_rate is the rate estimated upon each packet arrival.

The first part of Improved-TSW is the same as the TSW algorithm. The second part realizes the token bucket algorithm; when the average rate of the IN packet is above 1.33 * R_T, the algorithm checks whether there are enough tokens in the token bucket. If the response is yes the packet is kept as IN packet. Otherwise, the packet is remarked as OUT packet. Next we present an evaluation of the performance of the Improved-TSW and compare it with TSW and TCM.

Figure 7 shows that the average number of packets that are remarked from IN packet to OUT packet (at the boundary node Marker) using TSW, TCM and Improved-TSW respectively; the average number of packets changed from IN to OUT by using Improved-TSW algorithm is the lowest. TCM algorithm remarks more packets to OUT packets than TSW and Improved-TSW. Thus, TCM algorithm has the worst traffic performance. More IN packets are allowed to enter the AF service-capable network by using Improved-TSW algorithm; this means that more IN packets are protected by using Improved-TSW algorithm rather than TCM or TSW. Figure 2 shows that with Improved-TSW algorithm we get the best traffic performance, i.e., the amount of FTP traffic received by the receivers is the highest.

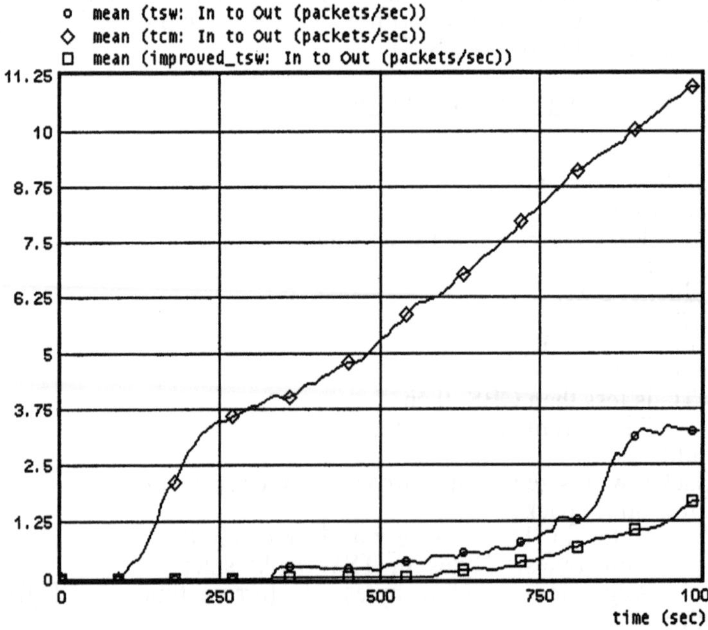

Figure 7 Comparison of three algorithms at Marker

5. CONCLUSIONS

In this paper, we compared two marking algorithms (TSW and TCM) that are currently used at the boundary nodes of Assured Forwarding networks. We proposed a new marking algorithm named Improved TSW marking algorithm that is designed by combining the TSW algorithm with a token bucket. We also proposed some other mechanisms to improve the performance of TCM algorithm.

This paper shows that selecting an appropriate marking algorithm is very important to AF service-capable networks. Our simulations show that

1. If we use TCM algorithm, a large amount of IN packets will be remarked to OUT at the boundary nodes of an AF service capable network;
2. If we use TSW algorithm or Improved-TSW algorithm traffic performance improves significantly; and
3. If we use Improved-TSW algorithm, we get better traffic performance (than using TSW) in our AF service-capable network with high burst and severe congestion at the bottleneck.

REFERENCES

[1] D. Black, S. Black, M. Carlson, E. Davies, Z. Wang, and W. Weiss, " An architecture for Differentiated Services", IETF RFC 2475, December 1998.

[2] D. Clark and W. Fang, " Explicit Allocation of Best Effort Delivery Service", IEEE/ACM Trans. on Networking, vol. 6, no. 4, pp. 362-373, August 1998.

[3] J. Heinanen, et al., " A Three color Marker", IETF Internet Draft, <draft-heinanen diffserv-tcm-01.txt>, February 1999.

[4] Juha Heinanen, Fred Baker, Walter Weiss and John Wroclawski, " Assured Forwarding PHB Group", IETF RFC 2597, June 1999.

[5] OPNET, http://www.mil3.com.

[6] K. Nichols, S. Blake, F. Baker and D. Black, "Definition of the Differentiated Services Field (DS Field) in the IPv4 and IPv6 Headers", RFC 2474, December 1998.

[7] S. Floyd and V. Jacobson, " Random Early Detection Gateways for Congestion Avoidance", IEEE/ACM Transactions on Networking, Vol. 1, No. 4, pp397 – 413, August 1993.

[8] Andrew S. Tanenbaum, "Computer Networks", 3 rd Ed, 1996, Prentice Hall, Inc., pp458 – 459.

[9] J. Ibanez and K. Nichols, " Preliminary Simulation Evaluation of an Assured Service", IETF Internet Draft <draft-ibanez-diffserv-assured-eval-00.txt>, August 1998.

Chapter 6

A Traffic Differentiation Method for IPv6 over ATM Networks

Jorge Sá Silva , Nuno Veiga , Sérgio Duarte and Fernando Boavida
{sasilva,nunov,sduarte,boavida}@dei.uc.pt
Centro de Informática e Sistemas da Universidade de Coimbra
Polo II - Pinhal de Marrocos
P-3030 Coimbra – PORTUGAL
Tel.: +351 39 790000 Fax.: +351 39 701266

Key words: Multimedia network management models and architectures

Abstract: At present, the number of Internet users continues to increase dramatically, accompanied by the large-scale introduction of new applications with multimedia capabilities. New users and applications are constantly demanding better performance from the Internet, if possible with some forms of quality of service (QoS) guarantees. Knowing that no single technology is able to provide such guarantees in the Internet environment – which, by definition is a multi-technology environment – the exploration of promising technologies capable of traffic differentiation that are applicable to parts of the Internet appears as a possible way forward. This paper addresses and explores the benefits of the Internet Protocol version 6 (IPv6) and Asynchronous Transfer Mode (ATM) integration in multimedia environments. A method for traffic differentiation in IPv6-over-ATM networks is proposed, based on the mapping of different IPv6 flows into the address resolution functions of ATM. The proposal is evaluated in terms of introduced overhead, by means of simulation studies, supported by an experimental platform and an IPv6-over-ATM simulator developed for this purpose.

1. INTRODUCTION

The current size of the Internet makes it very difficult to incorporate radical changes; yet, at least some changes are required in order to allow the support of multimedia and other QoS-requiring applications. Two of the technologies that, at the moment, have some potential to support multimedia applications are the ATM technology and the new version of the IP protocol, IPv6.

IPv6 was developed to overcome IPv4 limitations. The main improvements observed in IPv6 are concerned with address space – to overcome the current address exhaustion of the IPv4 – QoS support, self-configuration, security, efficient routing, multicast and anycast. On other hand, ATM is a proven technology, with good capabilities in terms of QoS support, widely deployed in Metropolitan Area Networks and Wide Area Networks. Nevertheless, when compared to other technologies, ATM is less deployed in local area network environments because of the complexity and the cost associated with it. On the other hand, ATM offers the flexibility to transport different services, respecting their needs and properties.

The combination of IPv6 and ATM places a set of new challenges, when compared to the IPv4 environment. As will be described in section 2, there are some key differences that require some extra functionality. In addition to identifying those differences, section 2 also summarises the IPv6-over-ATM "state of the art". Section 3 presents the 'Mars Extensions Developed for IPv6 over ATM' (MEDIA) project, in which scope the presented work was developed. In this section a new solution for the provision of quality of service in IPv6-over-ATM networks is also presented. Evaluation results concerning the proposed solution are presented and discussed in section 4. The conclusions and topics for further work are addressed in section 5.

2. IPV6-OVER-ATM NETWORKS

The main difference in IPv6 and ATM integration when compared to the IPv4 case is due to address resolution. In fact, in IPv4 environments using an overlay-model layer 3 to layer 2 address translation is made by auxiliary protocols like ARP or ATMARP. In contrast, IPv6 does this task by itself although, on other hand, it assumes a broadcast connectionless medium at layer 2 in order to implement Neighbour Discovery, Router Discovery and Address Configuration tasks [NAR96]. The apparent reason for this assumption is that IPv6 was developed assuming that the underlying network is an Ethernet.

In the Neighbour Discovery process, when the IPv6 layer needs to send an unicast message to some IPv6 destination address it looks for the corresponding layer 2 address in a translation table. If this layer 2 address is not found, the IPv6 layer sends a Neighbour Solicitation (NS) message to all in-link nodes. The node that has the corresponding IPv6 destination address receives the NS message and, if it accepts the connection, it establishes a point-to-point VCC and sends a Neighbour Advertisement (NA) to the calling node (Figure 1). When this is received, the source node can send the pending information, due to the bi-directionality of the established VCC.

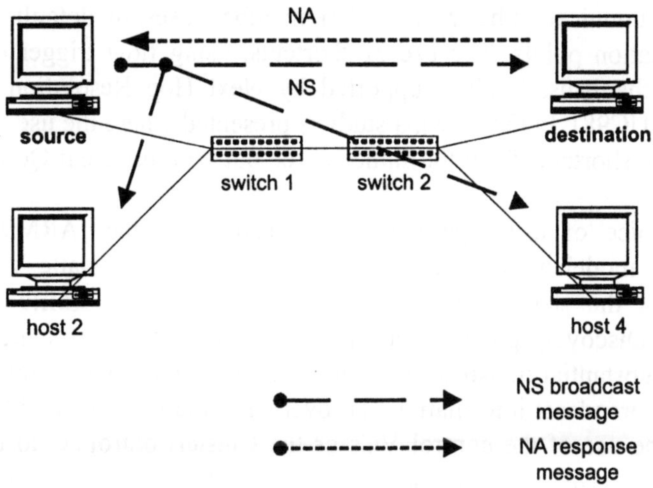

Figure 1. Neighbour Discovery process

The above mentioned process – originally based on the use of broadcast connectionless media – requires a new sublayer between the IPv6 layer and the ATM layer, hereafter called IPv6-over-ATM sublayer, that emulates the broadcast connectionless environment not natively provided by ATM networks.

Several approaches have been developed in recent years for the deployment of IPv6 over ATM. The Logical Link (LL) concept was introduced in [SCH96]. The LL is the ATM node set which can send and receive broadcast traffic. Each LL layer should emulate the layer 2 functionalities assumed by the IPv6 layer and, at the same time, should offer fault tolerance and redundant mechanisms. According to this proposal, each IPv6-over-ATM sublayer must capture all the broadcast traffic and send it to the Logical Link Server (LLS). This entity receives all broadcast traffic and sends it to all ATM nodes in the LL. There is no traffic differentiation and/or QoS mechanism present in this solution. In fact, this solution is totally based on the use of best-effort VCs. The LL concept can be seen as a "light-weight" solution, when compared to the MARS server approach that will be described next.

The Multicast Address Resolution Server (MARS) solution [ARM96] was originally proposed to translate IP multicast addresses to sets of ATM addresses, in multicast Classical IP environments [LAU94]. In [ARM99A] and [ARM99B], the use of the MARS approach by the IPv6-over-ATM sublayer is proposed. In a host, all the broadcast traffic sent by the IPv6 layer is captured by the IPv6-over-ATM sublayer and sent to the MARS server. After arrival at the server, the traffic is routed to all the cluster members.

The communication between off-link members uses, by default, *mrouters* as agglutination points. The use of shortcuts, using flow trigger or source trigger mechanisms, is also supported by Next Hop Resolution Protocol (NHRP) [LUC98]. In [TUR98], a study is presented where the use of a bit to differentiate shortcuts from the same source but with different QoS needs is proposed.

The choice of using permanent connections in the [ARM99A] and [ARM99B] model implies that the number of necessary static VCs is too high, as all the stations need to receive the broadcast traffic from the Neighbour Discovery process. On the other hand, the use of switched VCs implies a substantial transit of overhead messages whenever a station needs to transmit broadcast information. To overcome these problems [ARM96T] proposes the use of the control VCs or the ClusterControlVC to broadcast the Neighbour Discovery messages. It is a simple model, as it only needs few modifications to adapt the MARS server code.

[ATK96] proposes the introduction of an IPv6 layer version specific to the ATM technology. A new Neighbour Discovery process is presented, and it is also suggested the use of 6 tokens at the auto-configuration process. This proposal uses an NHRP extension to find the layer 2 address given the layer 3 address. Although the use of an ATM-specific IPv6 version could offer better performance results, it can be looked at as a major impediment for the IPv6 success, as it reduces its generality.

The MARS approach appears, then, as the most consistent approach for the provision of IPv6 over ATM networks. This was also recognised by IETF (Internet Engineering Task Force), which decided to adopt the MARS solution for IPv6-over-ATM environments. Nevertheless, the MARS model presents some limitations: it does not use some of the IPv6 benefits and there are some UNIv4 properties that are not sufficiently explored.

3. TRAFFIC DIFFERENTIATION

In the MEDIA project we are developing some extensions to the MARS approach, keeping full compatibility with the standard approach. The introduction of a new, original model would render difficult the success of the underlying ideas – no matter how interesting they would be – because of the wide acceptance of MARS and its support by IETF.

The MEDIA project intends to propose new extensions to CLIP (Classical IP) environments, that make use of the benefits inherent to IPv6 and UNIv4, not yet explored by other models. We are particularly interested in the use of additional QoS parameters in Neighbour Solicitation (NS) messages and in the IP multicast join requests – MARS_JOIN messages – for the support of traffic differentiation.

In unicast communications, we propose the use of the Flow Label Field or the Traffic Class Octet (TCO) IPv6 field in NS messages, using the Differentiated Services model [BER99], to transport the information concerning the requested QoS class. This value, when received by the destination node, is to be translated to the corresponding ATM QoS. The destination node then establishes a bi-directional point-to-point connection to the source node (figure 2).

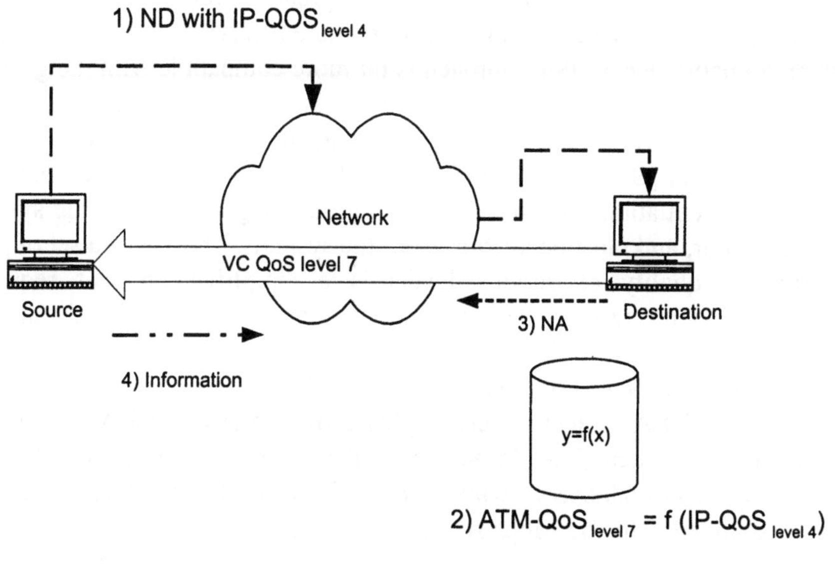

Figure 2. Proposed algorithm

As proposed in the Internet2 recommendations [INT97M], future services must support different QoS levels. When compared to the original process, the proposed solution respects QoS requirements.

As proposed by the Differentiated Services model, the QoS value at the IP layer is specified as an integer. The option between the use of the Flow

Label field and the TCO field depends on the number of bits necessary to distinguish the various QoS levels.

We propose the use of a single MARS server to all the ATM network. In our model the NS messages are directly sent to the destination nodes bypassing routers and the NHRP process, offering a more efficient solution when compared with [TUR98].

To not confuse the IDMR protocols and to not disbalance the multicast distribution tree, in case of multiple egress Mrouters, we use the single gateway principle, as proposed in EARTH [SMI97].

When, after receiving the NS message, the destination node cannot establish the requested VC with the required set of QoS parameters, it must ignore the NS message. The source node must, then, implement a timer that is activated whenever an NS message is sent. If the corresponding NA message does not arrive within a specified time interval, the source node must retransmit the NS message.

Although the majority of the existing approaches for QoS provision use the Integrated Services paradigm, Differentiated Services systems are now gaining importance, as this approach is far more compatible with the general Internet paradigm. The model presented in this paper is applicable when the ATM sub-network is part of a larger network respecting the Differentiated Services model [BER99]. The different QoS levels – that are at the basis of traffic differentiation – must be pre-configured, negotiated with the system administrator, and must be known by all the systems. The use of a set of pre-configured quality of service levels is a simplification that requires configuration by systems management functions. The use of a dynamical protocol to offer non-static QoS levels would complicate the proposed model and is outside the scope of the present proposal.

In the MEDIA project we are implementing an IPv6-over-ATM testbed using an experimental ATM network. The testbed is composed of two switches – a Fore200ASX switch and a 3COM 7000HD switch – and six hosts equipped with 155 Mbps ATM NICs (Figure 3).

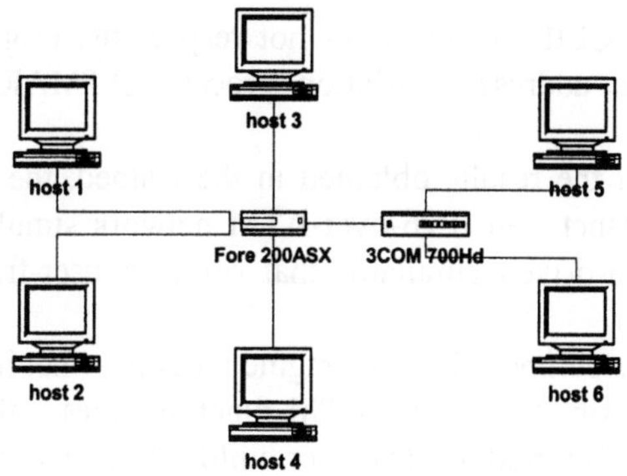

Figure 3. The MEDIA project testbed

We also developed the IPv6-over-ATM sublayer module and the MARS modules – the client module, the server module and the Multicast Server module. Each host has the modules presented in Figure 4.

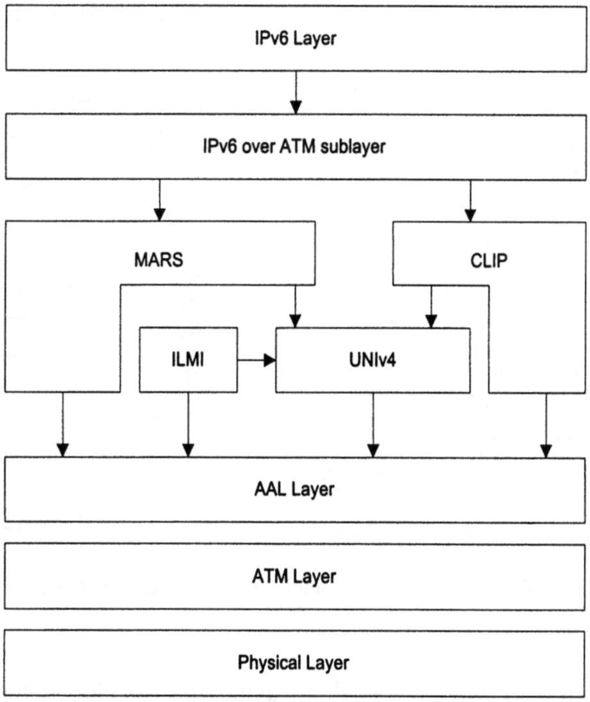

Figure 4. Protocol stack of IPv6 over ATM hosts in the MEDIA testbed

The developed CLIP version does not respect the original version. As IPv6 performs the address translation process, ATMARP is no longer present.

To complement the results obtained in the testbed, the MEDIA project team developed Sianet - an IPv6-over-ATM network simulator. Sianet is a C++ discrete event-driven simulator that offers a user-friendly graphical interface (Figure 6).

The simulator is composed of the engine module – that implements all of the previously described IPv6 and ATM functionalities – the user interface module, and the PNNI module. The user builds the network to be simulated using the graphical interface to specify hosts, switches and links, and selects a number of applications that run in each host (Figure 5). For each network element, different parameter sets can be specified.

Up to the present, we have developed some CBR and VBR applications. We plan to build some IPv6 native applications, as they have some different statistical properties when compared to IPv4 applications.]

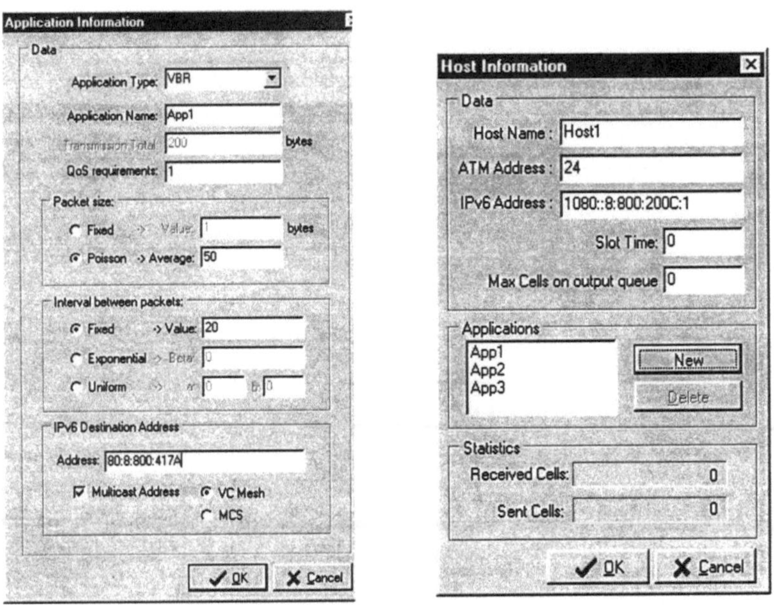

Figure 5. Examples of host and application parameter specifications

The validation of Sianet was done using the developed testbed. Also, some size and time overheads were obtained from the real platform.

In the context of the MEDIA project, Sianet is being used for the study of the proposed algorithm behaviour in scenarios other than the experimental testbed scenario.

In this project we are also studying and proposing traffic differentiation algorithms for multicast environments. We are particularly interested in the use of a new set of TLVs in MARS_REQUEST and MARS_JOIN messages to convey QoS requirements information in multicast communications. We are also developing some studies to evaluate the use of ATM multicast addresses in IPv6-over-ATM environments. ATM multicast addresses were introduced with UNIv4 but only for anycast environments. Although some PNNI modifications are necessary, the use of one-to-one mapping of IP multicast addresses and ATM multicast addresses is very powerful and can substantially reduce the number of MARS_MULTI messages.

4. OVERHEAD ANLYSIS

We made some simulation studies in order to quantify the worst possible overhead of the proposed solution in relation to the original ND process. When compared to the original process, the proposed algorithm should need some extra time to translate the IPv6 QoS requirements into QoS characteristics (more precisely, non best-effort VCs that correspond to the required QoS level) at the ATM layer. We assumed the use of hashing algorithms in the translation tables.

The following elementary mean time-overhead values were used for the simulation. These values were measured in the MEDIA testbed.

Table 1. Mean time overhead values

Overhead	Time (μs)
IPv6 layer: downwards direction	30
IPv6 layer: upwards direction	4
IPv6 layer: address table look up	0.6
IPv6 over ATM sublayer: unicast downwards direction	0.5
IPv6 over ATM sublayer: multicast downwards direction	384
IPv6 over ATM sublayer: upwards direction	0.5
CLIP layer: downwards direction	3
CLIP layer: upwards direction	2.2
MARS Server: processing of MARS_REQUEST	189
MARS Client: processing of MARS_MULTI	154
MARS Server: broadcasting of NS messages	120
MCS: broadcast messages	120
ATM layer: downwards direction	50
ATM layer: upwards direction	50

The measurement of each time value involved the identification of the initial and final points in the code of the operating system and driver. In these points the microtime() function was used to obtain the time values.

Table 2 presents some simulation results, in terms of the time needed to send 200 bytes of information from a host A to a host B, using a congested network composed of a varying number of intermediate switches (figure 6).

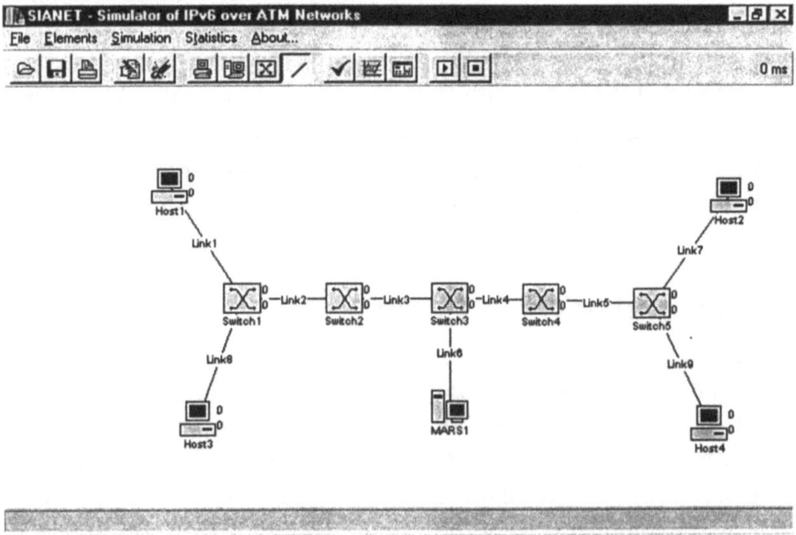

Figure 6. Sample configuration of a simulated network (5 switches)

The values obtained for each study were the result of the average of the values obtained in 10 simulation runs. The relative accuracy (for this number of simulations and for a confidence interval of 95 percent) varied between 1 and 4 percent. That is, it can be affirmed with 95 percent certainty that the real time average differs from the measured average in any given simulation run by a maximum of 1-4 percent.

Table 2. Time to send 200 bytes of unicast information

Number of switches	Time (μs)		
	original process	proposed process	overhead
1	20867	21447	580
2	41096	41688	592
3	61325	61929	604
4	81553	82170	617
5	101782	102412	630
10	202926	203618	692
20	405213	406030	817

As expected, the proposed solution does increase the time spent in the transmission of the information, but this increase is more or less constant (as the QoS mapping is done only once, prior to the establishment of the virtual channel) and negligible.

This same observation was confirmed by a second study. In this case, a network composed of 3 switches was used and the message lengths varied from 1 byte to 5000 bytes. Figure 7 presents the simulation results.

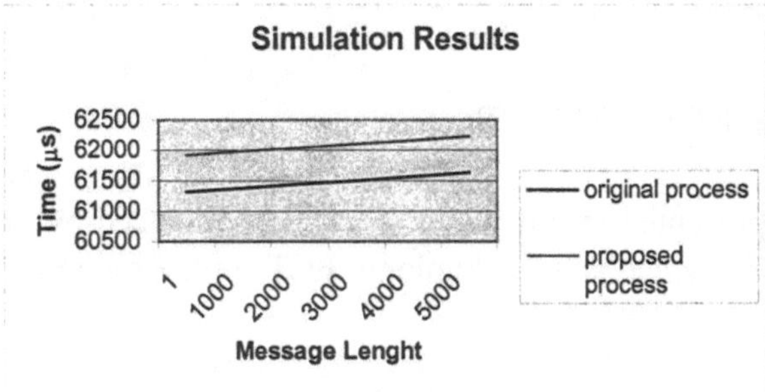

Figure 7. Transmission time as a function of message length, in IPv6-over-ATM unicast communication

As the message length increases, the relative overhead (that is, the ration overhead / message length) decreases, due to the fact that the introduced overhead affects only the establishment phase and is practically constant.

5. CONCLUSIONS

As multimedia and other QoS demanding applications become increasingly used in IP networks, traffic differentiation mechanisms take on a leading role in the efficient use of network resources. Several technologies have a good potential for the support of traffic with varying QoS requirements. Two of these – IPv6 and ATM – have been experiencing a growing acceptance over the last few years.

The existing proposals for the use of IPv6 in ATM networks were developed for best-effort environments and, thus, do not explore the traffic differentiation capabilities that are inherent to each of those technologies. The MEDIA project, briefly presented in the present paper, explores these capabilities, by proposing an approach to the mapping of IPv6 QoS requirements into VCs with appropriate quality of service characteristics. This mapping is done by an intermediate sublayer – the IPv6-over-ATM sublayer – using the standard, IETF-adopted address resolution mechanism.

Although the MEDIA project intends to analyse and propose new methods for unicast and multicast systems, this paper focused on the former case. After the presentation of the proposed QoS mapping solution, an analysis of the incurred overhead is performed, by means of simulation studies. The simulations showed that the proposed method introduces a small and almost constant overhead, which points to the fact that the proposed solution is feasible and that its inherent benefits have a negligible cost. Future developments of the proposed solution will address other IPv6 over ATM scenarios, like LAN emulation and MPOA.

ACKNOWLEDGEMENT

The work presented in this paper is partially financed by the Portuguese Foundation for Science and Technology, FCT, in the scope of the PRAXIS XXI program.

REFERENCES

[ARM96] G. Armitage "Support for Multicast over UNI 3.1 Networks based ATM Networks" RFC 2022 - 1996

[ARM96T] G. Armitage "Transient Neighbors for IPv6 over ATM" draft-armitage-ion-tn-01.txt - November 1996

[ARM99A] G. Armitage, P. Schulter, M. Jork and G. Harter "IPv6 over Non-Broadcast Multiple Access Networks" RFC 2491 - January 1999

[ARM99B] G. Armitage, P. Schulter and M. Jork "IPv6 over ATM Networks" RFC 2492 - January 1999

[ATK96] R. Atkinson "IPv6 over NBMA Networks" draft-ietf-ion-ipv6-nbma-00.txt - June 1996

[BER99] Y. Bernet et all "A Framework for Differentiated Services" draft-ietf-diffserv-framework-02.txt - February, 1999

[INT97M] Internet 2 "Internet Protocol Multicast Problem Statement", draft-bradner-multicast-problem-00.txt - September 1997

[LAU94] M. Laubach "Classical IP and ARP over ATM" RFC 1577 - January 1994

[LUC98] J. Luciani, D. Katz, Piscitello, B. Cole and N. Noraswamy "NBMA Next Hop Resolution Protocol (NHRP) " RFC 2461 - April 1998

[NAR96] T. Narten, E. Nordmark and W. Simpson "Neighbor Discovery for IP Version 6 (IPv6) " RFC 1970 - August 1996

[SCH96] P. Schulter "A Framework for IPv6 over ATM" draft-ietf-ion-ipv6atm-framework-00.txt

[SMI97] M. Smirnov GMD FOKUS "Efficient and Easy IP Multicast Routing Through ATM Networks", Proceedings of 7th IFIP Conference High Performance Networks 1997

[TUR98] S. Turner, and B. Staegemeir "Differentiated Services over Symetric NHRP Shortcuts" draft-turner-diff-nhrp-00.txt - March 1998

PART THREE

Multimedia Networks and Services

Chapter 7

Managing Distributed Conferences with ISABEL

T. Robles, H..Velayos, J. Quemada, T. de Miguel, S. Pavón, J. Salvachúa, G Huecas, E. Castro, M. Petit
Departamento de Ingeniería de Sistemas Telemáticos
Universidad Politécnica de Madrid
trobles@dit.upm.es

Key words: distributed conference, multimedia services, coordination and management

Abstract: ISABEL provides a platform for development of new multimedia services, including distributed conferences, with presentations and attendances located in rooms at different sites. ISABEL permits the definition of new services adapted to the specific requirements of each case. It builds its own communication network providing a uniform Quality of Service through the variety of interconnected sub-networks. Nevertheless, technical problems are not the only problems on this kind of events. Planning, management and realisation of a Distributed congress covering several continents are issues directly related with technology management and co-operation between persons and computer systems. This paper reviews technical and organisation problems involved in the realisation of a Distributed Congress and proposes an organisation model based on the ISABEL platform to deal with them.

1. INTRODUCTION

Many multimedia applications have been developed during the pass years to support co-operative work, tele-teaching and many other activities involving groups of persons distributed at different locations and communicated by means of telecommunication networks. Until now, most of this effort has been devoted to deal with technical problems: network design, network interconnection, robust applications, etc.). But we must also deal with problems related with the organisation of these activities as distributed events, involving sophisticated equipment and the users of the system.

Using a multimedia application we may organise and carry on a large number of distributed activities: videoconference, co-operative work, courses and congresses involving a large number of auditoriums located at different cities. This paper describes the material, the activities required for the organisation, and management of a distributed Congress using a service creation platform (ISABEL [5]).

World coverage implies the usage of long distance communication, with the problems inherent to protocol interconnection. It produces co-ordination problems like time differences, cultural diversity (language is one of them, but not the most important) and all the problems related with the organisation and management of spread groups.

Setting up a distributed event comprised: building a network able to support the interchange of multimedia and control information during the event; the use of a suitable tool for providing a service that fulfils the requirements of the event we are organising; and the use of tools and organisation methods to deal with planning, management and realisation of the event

2. ISABEL: A PLATFORM FOR SERVICE PROVISION

ISABEL is a CSCW application for co-operative working designed into the Broadband Programs of the European Union RACE and ACTS, that allows the creation of new remote collaborative services in the field of real time interactive interconnection of audiences [1]. The interconnection of remote audiences is understood as the communication of groups of persons located at auditoriums, or any other type of meeting rooms, using a terminal based on a CSCW application.

ISABEL has been used to build service platforms of a great variety of services with a high degree of inter-activity, such as tele-working sessions involving development groups of the same company; distributed courses among Universities at the same or different country; or the realisation of distributed conferences. ISABEL is based on the use of TCP/IP and has been optimised to work over heterogeneous sub-networks as ATM (based on ground or satellite links), Ethernet, ISDN, etc.

The application has been validated in a big number of large scale, local and international events. Some of the intentional more relevant events are ABC96, Global360 and IDC98. Among the local (Spanish) events, we must remark the organisations of TelecomI+D (Spanish Telecommunication Congress) during years 97, 98 and 99.

It has been specifically this intensive usage, even during the first phases of the development, that tuned ISABEL to specific requirements of the organisation of distributed events. It is important to remark that the final users of those events were persons not directly related with this technology.

Services based on ISABEL offer a set of facilities that make them especially suitable for developing global communication services in areas where the international co-operation is required:

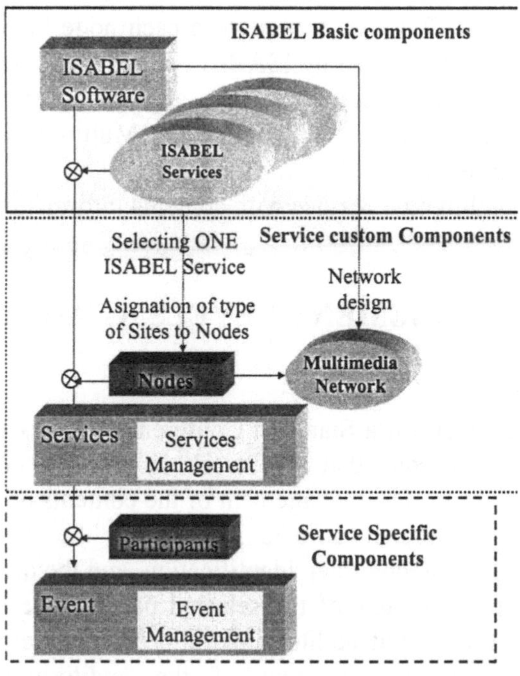

Figure 1 : Phases of the event organization with ISABEL

1. It offers a scalable architecture that may support the coverage of large
2. geographical areas and connect a large number of sites.
3. It offers the capacity of performing the global management of the event form a single point. It allows the design of strategies and service management in a very effective and user-friendly way.
4. It allows the definition of new services or the adaptation and tuning of already defined ones.
5. It works over heterogeneous networks, providing a uniform Quality of Service (QoS) by the use of a special adaptation agent (IROUTER [4]).

A distributed conference involves the participation of several auditoriums, from where speakers make presentations and people can attend at local or remote premises. The resulting event is a combination of a real time TV program with several studios, and a standard congress.

ISABEL provides a framework for the creation of new services(Figure 1) comprising three steps for the event realisation:

We start installing the ISABEL software and customising already defined services or defining new ones from scratch. Examples of ISABEL services are: tele-conference, tele-meeting- tele-working, etc.

1. During a second phase, once the service we want to provide has been identified, we set up a service that may be used for the realisation of several events. We need to assign to each node involved on the service one of the types defined by ISABEL, specifying which roles it will play during each mode. Combining the application characteristics and the participating nodes, we will define the Multimedia Network that will support the service.
2. Finally, combining a Service with contents information we carry events.

3. MANAGEMENT LEVELS AT A TELE-CONFERENCE

The management of a Standard Conference implies the co-ordination of two groups of persons that will work in parallel: contents groups and infrastructure group. The management of the contents of the conference is in charge of the conference diffusion, organising the call for papers, papers review and selection, chairmen identification, and the design of a program to organise the presentations of the selected papers. The infrastructure group will: reserve rooms and auditoriums, allocate secretariat of the congress, manage audio-visual equipment of the auditoriums, edition of the proceedings, management of inscriptions, selection of auxiliary personnel, publicity of the Conference, etc.

3.1 Communications Infrastructure (Network)

The provision of communication network with a sufficient Quality of Service (QoS from now) is today a research topic. The usage of a large variety of networks is required in world-wide coverage events, where different connection providers, network technologies and equipment suppliers must be involved. This scenario makes impossible the usage of a common QoS uniform for all of them. To deal with this problem ISABEL provides application level network software that helps to build its own communications network with end to end QoS.

This network is the essential element that connects the participating sites. The setting-up, configuration, maintenance, and operation of such network are complex problems by itself. When the Congress offers world-wide coverage, involving a large number of sites, configuration and management of such network becomes a critical problem. Increased by the fact that usually communication resources are only available in the last moment, by technical or economical reasons.

ISABEL as application is ready to work over a large variety of network technologies: optical fibber ATM links, satellite links, ISDN public lines (multiplexing several B channels), etc. All these sub-networks should be interconnected building an IP network. ISABEL will run over such IP network adapting multimedia flows to provided QoS.

3.1.1 Homogeneous network

The network is considered homogeneous when all ISABEL terminals can use the same network capabilities regarding available bandwidth, delay and packet drops. This is the simplest case, where it is possible to follow the conference with the same end user quality from every site. All terminals will share same multimedia flows (audio, video, slides,).

An important point is how multimedia flows are sent from sender to receiver. ISABEL supports several options depending on the network. If native IP multicast is supported by network, ISABEL will make use of this facility to replicate multimedia flows. Only one copy of each flow will be sent from senders to network, and network will deliver a copy to each receiver. If IP multicast is not available, ISABEL provides a Multipoint Control Unit called flowserver. Each sender will send a copy of each flow to the flowserver and this will deliver a copy to each receiver. During a conference, several flowservers can be connected building a distribution tree. Each ISABEL terminal must be connected to a flowserver.

Iit is possible to run a conference without flowservers. Each ISABEL terminal can act as MCU for other terminals. Only performance reasons impose a limit on the number of connected clients to each ISABEL terminal. The last option is to force each ISABEL terminal to work in a pure unicast basis, where each terminal will send a copy to each other participant site.

3.1.2 Heterogeneous network

In practice it is difficult to build world-wide coverage homogeneous networks. Each ISABEL terminal is connected at IP level but using different subnetwork technologies such as satellite links or fibber optic ATM circuits. We call this network heterogeneous cause each terminal will have a different network quality, different bandwidth, delay or packet drops. From the user point of view, this is perceived as different multimedia quality.

ISABEL is ready to work properly on these networks. Flowserver must be used in this case. Each ISABEL terminal will be connected to a flowserver using its particular network parameters, such as bandwidth available to the flowserver or expected delay. Flowserver will take care of bandwidth adaptation and proper buffering for each client. When bandwidth is severely reduced, flowserver can be requested to perform multimedia compression. This is possible because flowserver is an application level transport agent with lots of knowledge on multimedia flow contents and structure. This allows flowserver to receive multimedia flows, decoding and coding again using another algorithm on a per client basis.

Another useful function of flowservers is unicast to multicast flow conversion. When a terminal wants to join a multicast ISABEL session, but native IP multicast is not supported in its network, a flowserver must be configured to accept the unicast connection and feed multimedia flows in the multicast network and viceversa. One flowserver will be placed in the border of each multicast network and configured to exchange multicast traffic over a unicast connection.

3.2 Sites

An ISABEL Site is whatever place from where any body attends/participates on the conference. Those sites are also nodes of the Service. The Sites that will participate in the Conference may be classified as follows:

- **Interactive Site (IS):** from where attendants may interact (send and receive audio, video and data) with other Sites. IS may be classified for organisation and management:
  - **Main Site (MS):** a special IS, where there is audience, and the **speakers** will make presentations. It is highly recommended that they offer fail recovering facilities with back-up networks and equipments.
  - **Control Site (CS):** it is the most important site of the event, because it is in charge of the management of the whole conference, also performing remote control of the IS. The CS may be located at one of the MS, or may be an independent place without public nor speakers.

- **Watch Point (WP):** Sites that only receive multimedia flows sent by IS. They will not actively participate in the Conference, so they do not need a very well equipped Auditorium. They are simple passive receptors of information, and they do not interfere in the normal evolution of the Conference.

Relevance of the Sites is very important for the organisation and management of the Conference. The next levels of relevance has been identified:

1. The core of the Conference, composed by CS and MS. The disconnection of any IS allows other IS to continue with the conference. ISABEL allows that any IS assumes the role of CS, immediately and without disturbing the Conference development. Contingency plans must be scheduled to deal with problems at CS or MS.
2. The IS have also a big relevance, because they usually collect a larger number of attendants, that will participate in the turns of questions and discussions. Individual problems of the IS are not relevant problems for the Global Conference, but IS with frequent problems may jeopardise the development of the Conference.
3. Finally, Watchpoints are passive receptors of the distributed information. They do not interfere at any moment with the core of the Conference.

3.3 Contents

Contents provision and management using a distributed conference includes all the works requires for a standard congress, and tasks related with the distributed nature of the event: coordination of people placed in different sites and collection, transformation and distribution of the information material. To be distributed during the Conference

The Program Committee will produce Program with a number of sessions, presentations of each session and timetable of the Congress. In the case of a distributed Congress, the responsible of the Contents part of the Congress must coordinate with chairmen and speakers in order to decide from where they will participate in the Congress.

3.4 Services created with ISABEL

As shown in section 2 the first step in the definition of services using ISABEL, is to define the Types of Sites to be connected, the interaction modes and the roles that participants will play during each Interaction Mode. The definition of a new service is made by the creation of one **ISABEL activity**. Each activity is defined using a text file, that specifies the configuration of the ISABEL environment to behave in a specific way, and that defines, among another things:

- ISABEL modes (see 3.4.2.)
- Quality and format of sound.
- Control and interaction capabilities that each participant may use.
- Who may connect to the conference.
- Adaptation of the application to the underlaying network

The main facilities offered by ISABEL are:

- Management of the Distributed Conference from a Central Site.
- Distribution of documents and presentation from all the connected Sites.
- Diffusion of video and sound from any Site to all the connected.
- Independent selection and configuration of video and sound.
- Presentations with slides
- Blackboard for collaborative edition
- Display of the result of any program executed in any site.
- Control and management of questions and discussion panels.
- Definition and configuration of predefined ISABEL Modes.
- Definition of a detailed program and load it in the application, to facilitate the operation during the realisation of the Congress (see 3.3).

3.4.1 Conference Preparation and Control

Service preparation is fully based on Internet communications. Service operation features are centralised in a Web Server currently located at http://isabel.dit.upm.es. This server offers capabilities for: event description and creation, network definition and configuration, program creation and transformation, contents collection and transformation, test and rehearsal scheduling, coordination of people responsible for infrastructure and contents as well as of speakers and chairman.

ISABEL also allows the control of the whole conference from a single point using a graphical interface.

3.4.2 ISABEL Modes

The ISABEL Modes are screen layouts of different components (videos, slides, display sharing, etc). Typical ISABEL modes are:

1. **One video:** shows a single full screen video.
2. **Two videos:** the screen shows videos from two different auditoriums.
3. **Tree videos:** three videos from three different sites.

4. **Slides show:** the background of the screen shows the slides of the presentation provided by the speaker, and one video of the speaker is shown near one of the corners of the screen, usually the right top corner.
5. **Question mode:** the chairman selects one of the remote sites when questions arise, then the videos of the remote site and Speaker are shown on the Screen. A panel shows sites waiting for questions.
6. **Panel with 4 videos:** four videos from four different sites are displayed on the screen.
7. **Application Demo:** shows the execution of any application to all the IS connected to the conference by the use of a DISPLAY-SHARING.

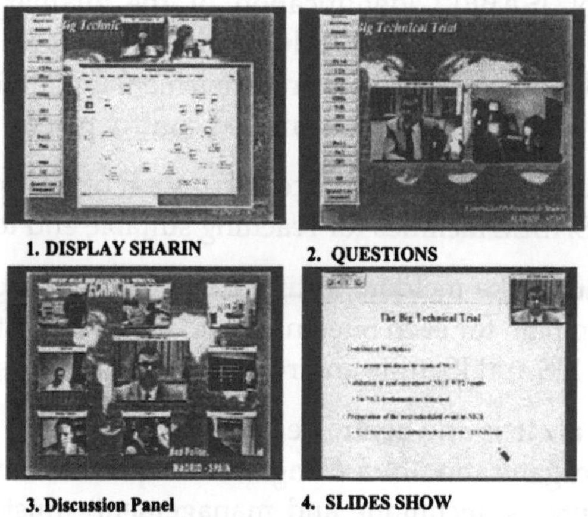

Figure 2. ISABEL modes at work

These are the most representative ISABEL modes, but new ones may be. ISABEL modes may be sequenced to produce a dynamic show, with a high degree of interaction and participation of sites. Figure 2 shows ISABEL modes at work.

4. PREPARATION OF THE DISTRIBUTED CONFERENCE

A Distributed Congress with ISABEL is organised in two parallel planes:

- **Contents plane,** which includes all the elements related to the information that participants will exchange during the Congress. Elements to take in consideration in this plane are relatively similar to the elements of a standard Congress.

- **Technical organisation plane,** which includes all the problems, related to the distributed organisation of the Congress. It includes the usage of a large communication network, a sophisticated application (ISABEL) and efficient coordination before and during the Congress

The persons involved in this group will be in charge of coordinating the Global Process, defining and maintaining links between the contents and the technical organisation planes. Their main objective is that the use of such a sophisticated technology is as transparent as possible.

The complexity of a Distributed Conference requires careful planning, and the realization of tests and rehearsals to minimise the number of incidents and problems during the Congress. The next elements must be planned and tested:

- **Set-up of the event on the web server.** It will provide basic support to: configuration and operation of the Communication network; installation and operation of ISABEL; identification of the auxiliary equipment required for the Auditoriums; scheduling and execution of tests and rehearsals; coordination of speakers and chairmen. E-mail list is used for the coordination of the previously identified control planes.

- **Network definition.** Allocation and configuration of network resources combined with ISABEL facilities for reaching suitable end to end QoS.

- **Testing Plan.** Must take include: testing backbone and access network, individual testing plan for each Site, interconnection and interoperability among MS, and MS and IS, tests and rehearsals of specific sessions

- **Detailed program.** It is the central element for defining the contents, scheduling and organisation of a Congress. All persons involved in the organisation (contents, technique and management) must have a deep knowledge of the detailed program to ensure a good coordination

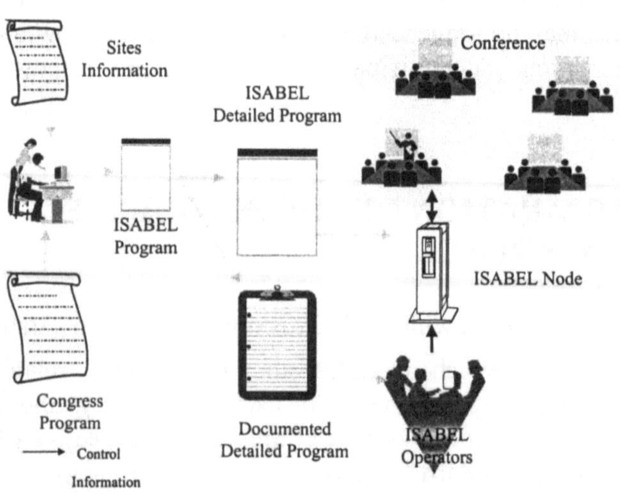

Figure 3. Program Management

The main input for the contents organisation is the program of the congress. This program must be converted into a standard electronic format to be processed. This standard format also includes information about the location of speakers and chairman during the sessions. It generates two outputs: the detailed script for ISABEL and a textual script to be used by the ISABEL operators during the event. The process is summarized in Figure 3.

5. REALISATION OF THE DISTRIBUTED CONFERENCE

The first element to define is the communication infrastructure that the organisation will use.

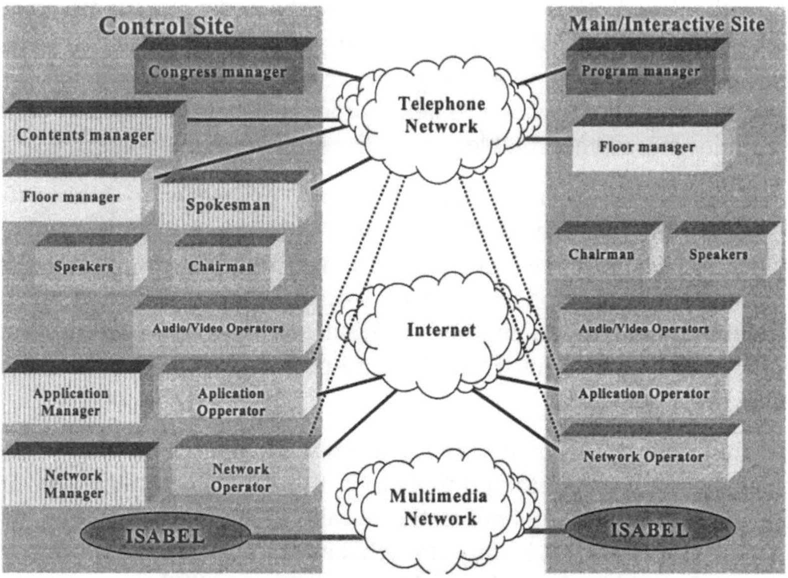

Figure 4. Communications during the Conference

Figure 4 describes the main communication mediums that will be used during the conference:

- The Communication Multimedia Network that connects the Sites that will support the Conference.
- Both the Application and Network Operators will use the standard Internet communication mechanisms. The use of IRC (Internet Relay Chat) has proved to be very useful as a private communication mechanism for co-ordination, notification of incidences, etc. If Internet connection fails, telephone will be used to connect the MS with the CS.

- For large and complex events, an audio-conference service has been successfully used for co-ordination sessions among the Congress Manager, the Program Managers of each Site, Spokesman, and eventually the Contents Manager.

A graphical interface allows to manually configure the application in real time during the event to fix last minute modifications to script.

5.1 Monitoring

In order to reduce the number of incidences perceived by the attendants during the Congress, tree elements must be continuously monitored:

- **The Network:** to anticipate any problem that may impact the QoS that allows the Application to offer a good quality to the attendants. In this kind of situations it is always better to apply preventive actions such as introducing breaks or modifying the Program while the network is fixed. Waiting till network problems produce a big malfunction of the application it is not acceptable.
- **The Application:** to avoid local problems propagate to other sites. Special attention must be paid to the sites with repeated and continuous problems. Sites with frequent problems will degrade dramatically the behaviour of the application, and jeopardise the evolution of the Distributed Conference. In this situation the Congress Manager may decide to disconnect some Sites during a period of time, until they have solved the problems.
- **The Contents:** to solve problems, usually considered as minor ones during standard congress, but that are in this context specially relevant. Like the late arrival of one of the speakers, the large duration of a welcome speech, etc,. It is important that all the sites receive precise and in time information about this incidences, its causes (if they are known), duration (if we may foresee), and which will be the next step.

6. CONCLUSIONS

The organisation and development of a Distributed Congress using Multimedia Application and Networks suitable for supporting multimedia traffic requires a wide usage of human and material resources. Solved, almost in part, the technological problems related with network bandwidth, and the creation and usage of new and powerful tools for services creation (ISABEL), organisation problems start to be the main problem. Facilities and tools for helping in the management of such services and event must be part of the Applications that will be used in the organisation of this kind of Distributed events. These requirements affect to the network management, the exchange of multimedia flows or whatever element that is part of the organisation. The information and their management must be considered an important element in every management organisation, even more if there are involved groups of persons spread along the world. A good information management, using the most advances technologies and tools available, will be one of the keys to success.

A good preparation, including tests and rehearsals of the real situations that will happen during the event, is the best way to ensure that every body will know what to do during the real event. Motivation and implication of persons involved in the organisation, including the participants in the contents part, also the speakers, is an essential element to have a successful Congress. The organisation must be supported by a wide communication platform that allows a fast decision agreement with the best and large information possible during the duration of the event.

Unexpected problems must also be foreseen. We must define back-up plans and alternatives for dealing with big problems as general failure of Communication Networks, unavailability of one or more MS, etc.

In this context becomes necessary the development of applications specially dedicated to the organisation, management and coronations of the events. In a short time, we plan to produce a workflow tool that will be integrated with ISABEL supporting the management of events.

REFERENCES

[1] J. Quemada, T.P. de Miguel A. Azcorra, S. Pavón, J. Salvachúa, M. Petit, D. Larrabeiti, T. Robles, G. Huecas, D. Rodríguez, F. Echevarrieta, E. Castro. *Teleservice creation with ISABEL in heterogeneous network environments.* Interoperability of Networks for Interoperable Services, Ottawa, July 1998.

[2] D. Larrabeiti, M.C. Agúndez, A. Azcorra, C. García, J. Quemada, T. De Miguel and M. Petit, *Towards the integrated configuration and management of multicast teleconferences based on IP over ATM.*HP Openview University Association Workshop, Madrid, April1997.

[3] M. Petit, T.P. de Miguel. *Distribution of Multimedia Information Experiences over Broadband Networks.* EUNICE '96 Summer School. Lausane, September 1996.

[4] M. Petit, T.P. de Miguel. *Flexible Support for Advanced CSCW Conferences: the ISABEL irouter..* PROMS'96: Protocols for Multimedia Systems, Madrid, Oct. 1996.

[5] J. Quemada, T.P. de Miguel, A. Azcorra, S. Pavón, J. Salvachúa, M. Petit, D. Larrabeiti T. Robles, G. Huecas *ISABEL: A CSCW application for the distribution of events..* COST 237 Workshop on Multimedia Networks and Systems, Barcelona, November 1996.

Chapter 8

Benchmarking Issues of a Distributed Multimedia Database Management Service

André Luís Vasconcelos Coelho and Ivan Luiz Marques Ricarte
DCA – FEEC – UNICAMP, PO Box 6101
Campinas, SP, Brazil, Zip-code 13083-970

Key words: Multimedia middleware service; distributed persistence management layer; CORBA; benchmarking.

Abstract: One of the most compelling yet still not properly accomplished tasks inside the distributed multimedia database arena is that of suitably fulfilling the disparate temporal requirements commonly imposed by the different types of persistent multimedia information. In such regard, an object-based multimedia management layer has already been conceived and implemented as a CORBA service towards the seamless integration of multimedia databases with "net-hidden" media-dedicated storage repositories. The goal of this paper is to present some quantitative results obtained through measurement sessions over a developed prototype of this distributed service in order to qualitatively assess the adequacy of its architectural proposal.

1. INTRODUCTION

One of the core issues in a distributed environment is to seamlessly integrate in a unified management framework a range of different specialised services. This is the typical case when multimedia functionalities should take place in. Such integration assumes an even more significant role if one strives to follow on the philosophy behind the openness precept; in this realm, integration has turned to be called interoperation and is the main target to be tackled by middleware tools [1].

Particularly within the distributed multimedia database systems (DMMDBMS) arena, there is one key factor regarding the manipulation of multimedia information that does still lack newer proposals for more suited architectural middleware solutions: How to endow persistence to multimedia objects (MOs). For such support, the approach followed by the Multiware Database platform [2] was to maintain within the same distributed multimedia environment at least one OODBMS and many media-dedicated repository servers. In this scenario, the OODBMS keeps in charge of the high-level media objects inter-relationship specifications while the servers should be specifically configured to take care of the media objects' active contents and streams retaining structural metadata information about their particular codification format.

In order to co-ordinate and control the access to these possibly distinct repositories, a distributed component-based mechanism had to be devised. The result was the design and implementation of a multimedia object persistence management layer called SGPOM [3], whose functional and operational features were motivated by proposals for handling media streams in Open Distributed Processing [4] and by the interoperability and portability trends advocated by the OMG CORBA standardised specifications (in particular, the CORBA Persistent Object Service [5]).

Aiming at the evaluation of this management service under real-work conditions, a prospective archetype was implemented in Java. Despite a few isolated initiatives towards the performance characterisation of some multimedia middleware services [6] or towards the adoption and launching of some benchmarking rules [7, 8], there is still much work to do until the stable maturation of assessment criteria to be employed over this new course of distributed technology. Hence, the contribution of this paper is to aid efforts in such endeavour as it tries to fill part of this gap when of the SGPOM performance evaluation. The expected effect is to set forth some initial results, which could serve as a basis for comparison with other CORBA-based distributed multimedia database systems [9].

The document is structured as follows. Section 2 is devoted to the general architectural overview of the proposed distributed persistence management layer, focusing at the inter-relationship between its (sub)components. Section 3 presents quantitative results obtained from an SGPOM prototypical implementation. These results are displayed by means of graphics and tables centred around some performance measures and parameters, confirming the expected service behaviour. Finally, Section 4 addresses some final considerations and points out some possible extension works.

2. SGPOM ARCHITECTURAL DESCRIPTION

The large-scale distribution of media-dedicated repositories seems to be a promising strategy for achieving distributed persistence for multimedia components, as it allows for a gamut of new functionalities: load balancing, data replication and sharing for fault-tolerance, increased trustability and availability, and autonomy and scalability to all repositories. SGPOM assumes the liability for the co-operative integration of these repository servers by orchestrating their activities and by controlling the access to them. It is a middleware layer with public interfaces, allowing applications to register, store, recover, and remove[1] their multimedia components' active contents in accordance with their temporal constraints. Its architecture is ruled by managers which integrate an expandable hierarchy based on CORBA POS, bringing transparency functionalities to the Multiware Database clients.

Fig. 1 depicts the SGPOM hierarchical framework, emphasising the inter-relationships between its management elements. Following POS philosophy, SGPOM presents itself as a stratified service composed by a rank of specialised components that inter-operate across an ORB substratum towards the accomplishment of the management and access activities. These entities are accessible through the set of public IDL interfaces of their constituent subcomponents and are each mapped to a level in the structure.

SGPOM enhances the capabilities of the POS router element, the Persistent Object Manager (POM), by providing an extensible set of new specialised managers, each related to a particular media type/subtype. SGPOM_PO is the unique visible component from the perspective of an SGPOM client; it acts as a higher level interface preserving the semantics of a typical persistent object. The set of managers (SGPOM_OMs) is headed by the LEADER, which behaves as the only CORBA server to which the I/O operations sent by the SGPOM_PO are forwarded. As every management component, the LEADER maintains information concerning its predecessors, the media-type managers (TYPEs). These, in turn, compose a level of the hierarchy representing existent media types, such as text, audio, and video. A manager of such class has associated with it a team of SUBTYPEs, each regarding a particular coding format (subtype) for a given media. The lowest layer is associated to the SGPOM_DSs, specialised "drives" responsible for interfacing with the media-dedicated repositories (Datastores). A SUBTYPE keeps with it a symbolic link to one or more SGPOM_DSs, each supporting a particular kind of Protocol. Because it converts the high-level I/O operations to the Datastore API, an SGPOM_DS resembles a database adaptor. Finally, the GENERIC component is

[1] These will be referred as I/O operations.

an entity viewed both as an adaptor and a Datastore to which I/O operations relating to *raw data*[2] should be forwarded.

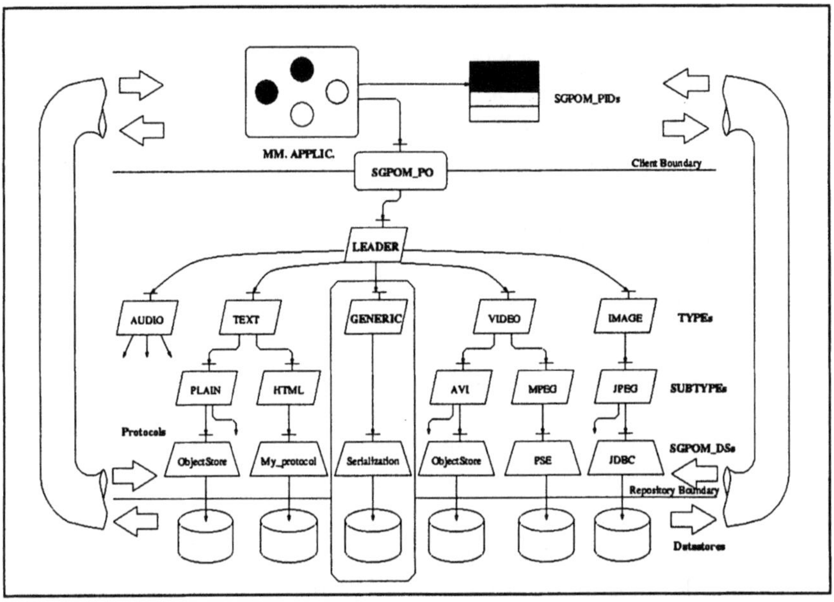

Figure 1. SGPOM framework of interoperable components

For SGPOM benchmarking purposes, the following concise list of design criteria [3] were considered. The basic I/O operations are divided into *management operations*, `Register()` and `Delete()`, which do not entail the direct handling of the MOs, and *access operations*, `Comm_way_Store()`, `Store()` and `Restore()`, which imply the deployment of an adequate communication middleware service to keep in charge of the media contents conveyance.

Each kind of entity in the SGPOM architecture—SGPOM_PO, SGPOM_OMs and SGPOM_DSs—actually abstracts one layer of the hierarchy encompassing a set of specialised elements. These elements are assigned to IDL interfaces defining four types of possible subcomponents, Factory, Controller, Central, and Child (Fig. 2). A Factory is a CORBA object in constant execution (persistent activation mode) responsible for the (re)creation of the instances of the other members of its layer—except for the Child objects—inside the same process. In turn, the Controller keeps in charge of local monitoring tasks. It constantly collects some performance values that quantitatively assess how its associated Central subcomponent has been servicing its addressed requests.

[2] Media contents that could not be either classified in pairs of types and subtypes or for which there does not exist a media repository available to handle their codification features.

The Central assumes the name of the layer it represents—SGPOM_*XX*, where *XX* is one of PO, OM or DS—, being conceived as its most important element. Each element in this class is only related to an SGPOM instance and may co-exist, in the same OS process, with peers representing other instances. Child objects were devised for scalability purposes: To support the forwarding of multiple concurrent I/O requests throughout the hierarchy. At each tier, there can exist an ever-growing pool of threads relating to Child subcomponents.

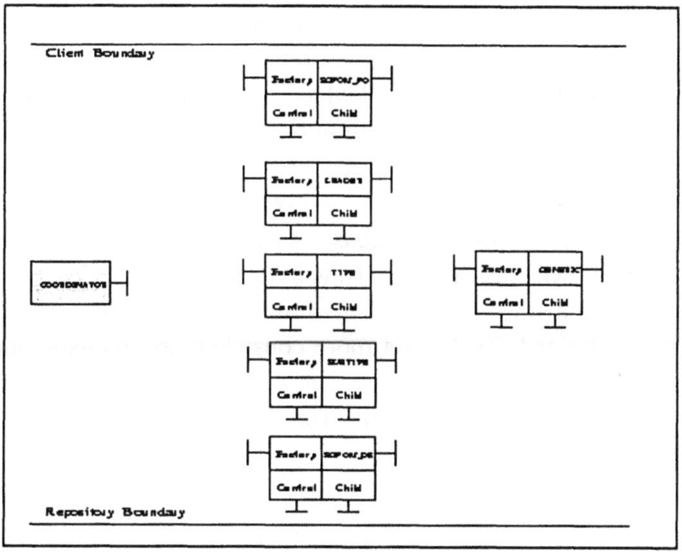

Figure 2. SGPOM subcomponents and their Coordinator

Within the SGPOM context, the deployment of a detached CORBA object, located outside the layered hierarchy and entitled as the Coordinator, was the chosen approach towards the (sub)components orchestration. Basically, this entity accomplishes the dynamic launching of Central and Controller instances, the periodic retrieval and assessment of performance parameters associated to each layer, and the regular checkpointing of all Central objects, forcing them to maintain their own executing state inside the GENERIC repository.

Aiming at a fair support to the concurrent access to the dispersed multimedia data, the POS POM functionalities was expanded, introducing new layers in the top-down structure. The result was the inclusion of groups of TYPEs and SUBTYPEs to represent, respectively, the MOs' media types and code formats, following the MIME standardised classification scheme. Besides, the idea of explicitly detaching the media streaming control and communication infrastructures [4] motivated the conception of an abstract container (named as Comm_way) to encapsulate a broad range of existing transport protocols and services. The goal was to bring up an alternative way to surpass the performance constraints imposed by conventional ORBs [6].

3. PERFORMANCE EVALUATION

For the accomplishment of the benchmarking sequence, an application scenario (*multimedia playback*) was envisaged, regarding the real-time video-audio streams retrieval and exhibition. The applet of Fig.3 was employed for the performance measurement collections (I/O requests invocation times) and to emulate a playback application. With this interface, one can opt to test a given I/O operation or a continuous series of them all, comprehending the MO's typical lifecycle: registry, storage, recovery, and deletion. With a URL to a media archive, an MO can be created in consonance with a high-level database schema, encapsulating the media data-stream as its own class attribute (raw byte array).

Fig. 4 presents the physical environment where the performance measures were attained. To keep a more consistent and realistic analysis, a heterogeneous testbed composed by machines with differentiated resources was devised. The benchmarking data were obtained in many sessions and under uniform conditions.

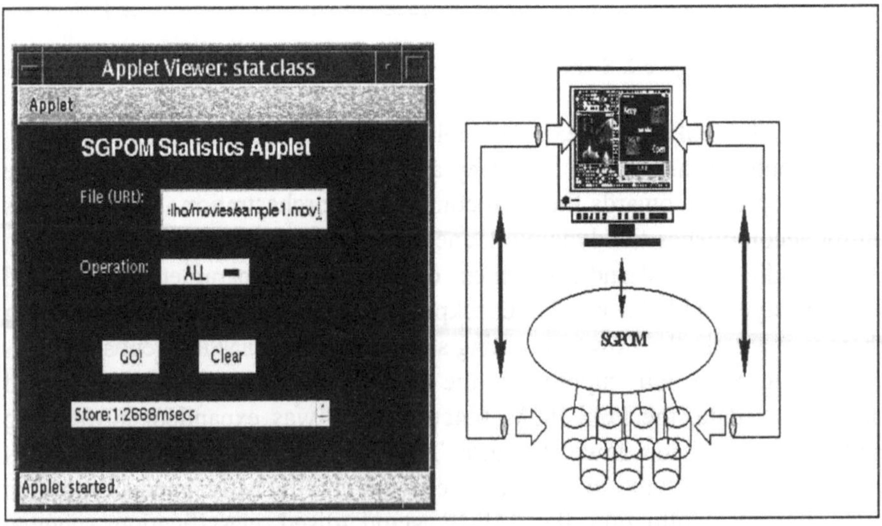

Figure 3. The statistics interface for benchmarking and the SGPOM playback scenario.

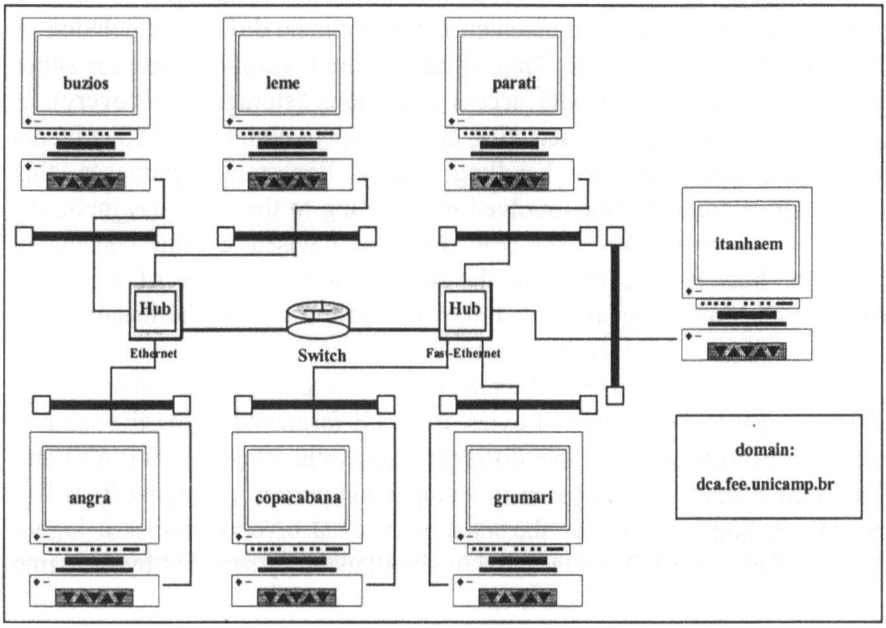

Figure 4. SGPOM testbed.

3.1 Metrics

For the service evaluation, the following group of *measures* was defined, focusing on the temporal aspects of the I/O requests attendance. The *Invocation Time* for a simple request means the elapsed time interval between two actions: The first comprises the event generated when clicking at the statistics interface "GO!" button, and the latter refers to the same interface receiving the request answer—for both management operations and for the storage one, this answer translates to a simple "success or failure" acknowledgement, while for the recovery sequence it consists of the MO effective restore. In case of a complete cycle of requests (registry–storage–recovery–removal of the MO), the start-up time of one operation (exception made to the subscription one) will be delimited by the final time of the previous one.

For comparison purposes, the focus here will centre upon the arithmetic average of those intervals (*Average Invocation Time*[3]). The *Management Time* refers to the Invocation Time for either one of the management

[3] In this section, $\underline{T}_x$ will represent such measure.

operations, i.e., registry or removal. With this measure, it can be inferred the time uniquely spent by the SGPOM layer components (that is, owing exclusively to the service arrangement), as there is no direct manipulation of the MO contents. The *Access Time* alludes to the Invocation Time for either one of the two types of MO access operations (storage or recovery). It provides a direct means for detecting the time spent both in the MO contents conveyance and repository handling phases. For storage purposes, this comprehends the total time involved in attending to the `Comm_way_Store()` and `Store()` operations. The *Efficiency* is devised as the percentage difference between the unit and the ratio between the sum of the mean management times (registry plus removal) and the sum of the mean access times (storage plus recovery) to a same MO.

Some parameters influenced on the data set obtained in the test sequence. *MO size* regards the number of bytes of a well-formatted media file sample. For a prospective analysis, five different sized archives (in MPEG, AVI and Quicktime formats) were chosen for comparison purposes, ranging from 1 to 25 MB. A higher value for the *number of Child objects* implies a higher number of possible I/O requests being simultaneously serviced by the same SGPOM component. At the same time, a larger internal data structure of the Central object is required to keep the relative Object References (ObjRef) to its Child related threads. Moreover, the time spent in the search for a free Child for the attendance of a new request would be increased. For the the measures presented here, a constant number for this parameter (20) was established with respect to all service component layers. The *number of clients* refers to the number of applications performing concurrent I/O requests over the same SGPOM instance. In this analysis, a unique statistic applet is used.

Unless for the SGPOM_DS, which directly handles the OM contents, the time spent in each tier can be deduced from the sequence of actions of Fig. 5: (*i*) A Child from the previous layer sends a `Create()` request to SGPOM_XX; (*ii*) SGPOM_XX looks up the table for the first available Child; (*iii*) SGPOM_XX returns the chosen Child's ObjRef; (*iv*) The Child of the previous layer sends an I/O request to the chosen peer; (*v*) This one restarts the cycle described above and then enters in a "blocked" state awaiting the forthcoming result; and (*vi*) the request result is propagated and the Child is relieved.

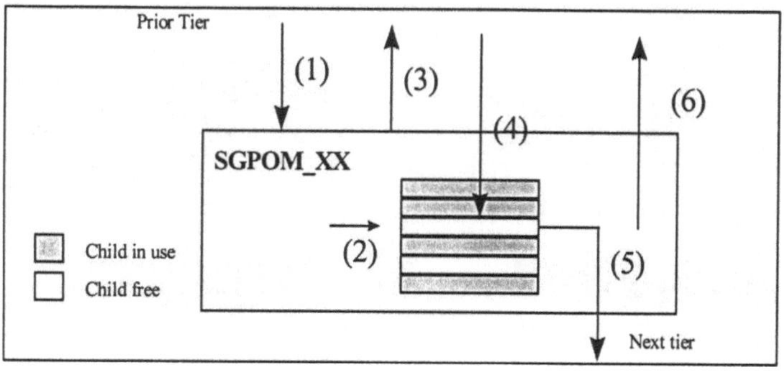

Figure 5. Action sequence chart concerning the time spent in each SGPOM layer.

3.2 Results

For the benchmarking accomplishment, IONA's OrbixWeb [10] was chosen as a supportive toolkit as it provides a robust IIOP-based ORB implementation with some additional facilities (such as *filters* and *loaders*), being compliant with the standardised Java-IDL mapping. Java Sockets were employed as the default media streaming communication infrastructure and Java Serialization as the default MO storage and codification protocol.

Fig. 6 (Layout #1) brings values for the "Invocation Time" measure regarding the four types of I/O operations—`Register()`, `Comm_way_Store()` + `Store()`, `Restore()` and `Delete()`—and the service *Basic Configuration*, set by an instance of SGPOM_PO, a LEADER and a GENERIC, each in a different host. This is done for the five sample archives ("MO Size") through complete cycles of I/O requests. The average values were calculated over a series of ten execution cycles, and the MO transmission, for storage/recovery purposes, is done exclusively over the Fast-Ethernet subnet via socket channels between similar machines (UltraSparc 256MB). Table 1 shows "Efficiency" values calculated by applying the ratio:

$$Efficiency = 1 - \frac{Tm_{management}}{Tm_{access}}, \quad (1)$$

where $Tm_{management}$ is the arithmetic average between the mean registry invocation time and the mean deletion invocation time, whereas Tm_{access} is the average between the mean storage invocation time and the mean restore invocation time.

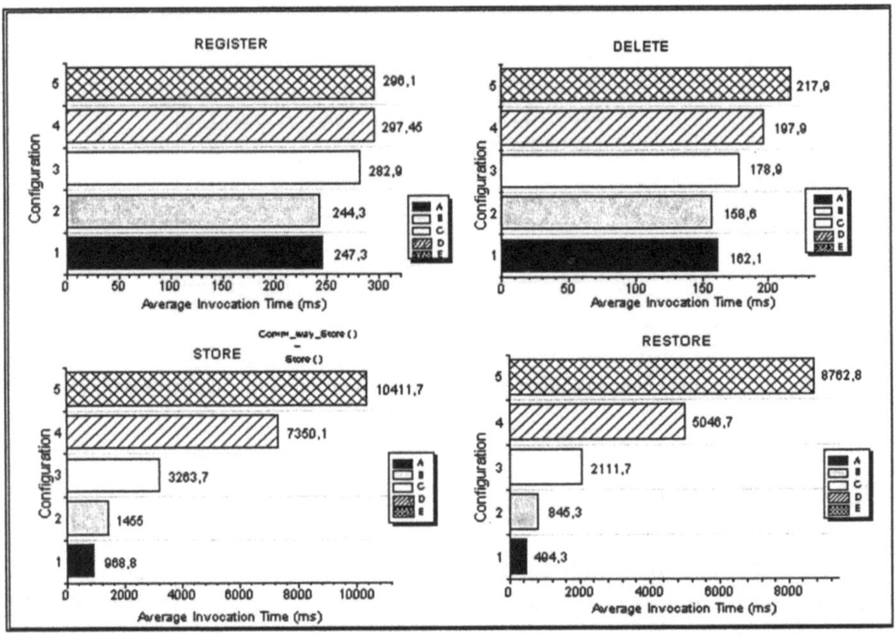

Figure 6. "Average Invocation Time" for the four I/O request types (Basic Configuration).

Table 1. "Efficiency" measure for the data set of Layout #1.

Archive	$Tm_{management}$ (ms)	Tm_{access} (ms)	*Efficiency* (%)
Sample1 (930 KB)	204.68	731.55	72.02
Sample2 (2140 KB)	201.43	1150.15	82.49
Sample3 (6.5 MB)	230.9	2687.7	91.41
Sample4 (15.5 MB)	247.67	6198.4	96.00
Sample5 (23.17 MB)	256.99	9587.25	97.32
Average	**228.33**	**4071.01**	**94.39**

The management operations, despite small differences, present a uniform (deterministic) temporal behaviour, centring around the 230 ms bounds. On the other hand, the access requests demonstrate, as expected, a continuous growth for the invocation times, in proportion to "MO Size". Large values for this parameter implies large "Efficiency" results, as the proportional cost of time due exclusively to the SGPOM layers' forwarding activities will be less significant to the total amount. Consequently, the "Efficiency" measure turns to be a good cost/benefits indicator for SGPOM deployment.

A detailed report of the "Average Invocation Time" was undertaken by separating the lots of time in agreement with the following independent activities: MO transmission (average time spent with sockets, $\underline{T}_{socket}$), MO management and handling at the repository ($\underline{T}_{Datastore}$) and MO memory encapsulation ($\underline{T}_{memory}$). By this way, the temporal cost associated exclusively to the SGPOM Basic Configuration activities ($\underline{T}_{SGPOM}$) could be deduced, both for the access and management operations. In this respect, Layout #2 (Fig. 7) depicts, in detail, the percentage quota relative to each of the activities for the "Average Invocation Time" measure, regarding the MOs management requests. The $\underline{T}_{SGPOM}$ can be inferred, in this case, by the following simple algebraic equation:

$$\underline{T}_{SGPOM} = \underline{T}_{Total} - \underline{T}_{Datastore}, \quad (2)$$

where $\underline{T}_{Total}$ are the values of "Average Invocation Time" in Layout #1.

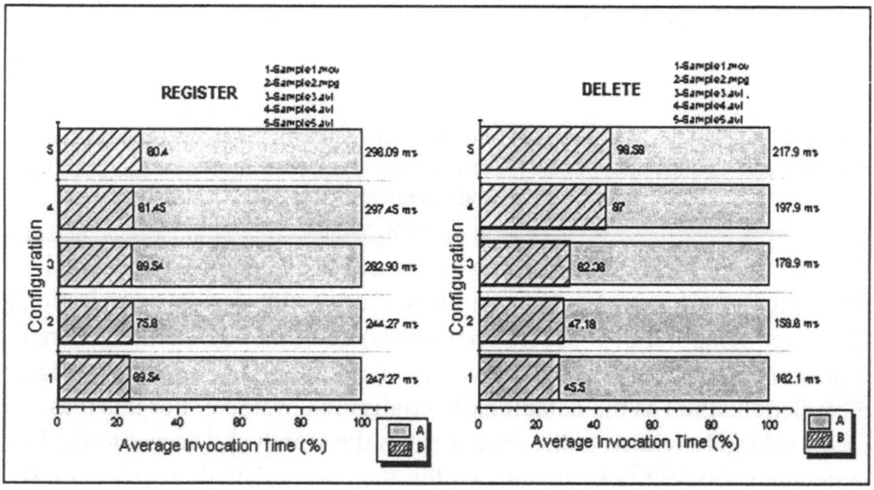

Figure 7. Detail of the "Average Invocation Time" measure for the management operations, in SGPOM Basic Configuration, where (A) indicates the $\underline{T}_{Total}$ and (B) indicates the $\underline{T}_{Datastore}$.

Layout #3 (Fig. 8) follows the guideline stated in the previous paragraph. However, here, $\underline{T}_{SGPOM}$ can be inferred, for the MOs storage and recovery requests, from their respective associated expressions:

$$\underline{T}_{SGPOM} = \underline{T}_{Total} - \underline{T}_{Datastore} - \underline{T}_{socket} - \underline{T}_{memory}, \quad (3)$$

$$\underline{T}_{SGPOM} = \underline{T}_{Total} - \underline{T}_{Datastore} - \underline{T}_{socket}. \quad (4)$$

$\underline{T}_{memory}$ concerns with the execution time spent in reading (for the memory) the flow of media composing the MO, giving the proper URL to the media archive. For that reason, it is only examined within the storage activities.

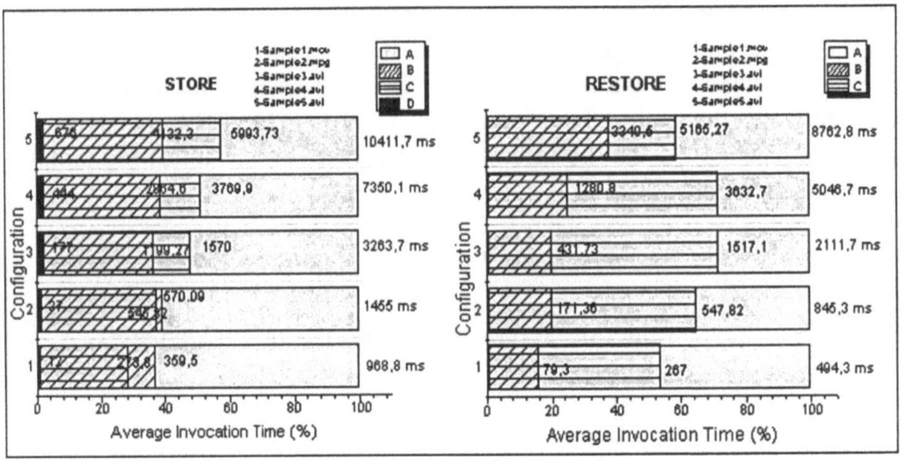

Figure 8. Detail of the "Average Invocation Time" measure for the access operations, in SGPOM Basic Configuration, where (A) indicates the $\underline{T}_{Total}$, (B) indicates the $\underline{T}_{Datastore}$, (C) refers to $\underline{T}_{socket}$, and (D) refers to $\underline{T}_{memory}$.

At the time of the benchmarking sessions, the availability of a complete implementation of a media-dedicated repository (with design optimisations and assured temporal disk access restrictions) could not be fulfilled. Hence, the network file system was deployed as the default Datastore and the Java Object Serialization as the default data serialisation protocol. For this reason, $\underline{T}_{Datastore}$ is largely related to the time spent in the search and access to the file that contains the MO raw contents within the appropriate directory (usually remote); such access is made possible through NFS support.

For registry activities, the $\underline{T}_{memory}$ temporal contribution stays soundly constant, as it is limited to the creation of a null content file and of an identification string (SGPOM_PID). For the removal task, the quote due to $\underline{T}_{Datastore}$ grows as the handled MO volume increases, too, which agrees with what is expected: The time spent in the physical removal of the object bytes in the disk will be higher. $\underline{T}_{Datastore}$ as much as $\underline{T}_{socket}$ are the most contributors to the storage and recovery temporal costs (MOs direct handling) whereas the time strictly spent with the SGPOM hierarchy stayed within the limits of a uniform range (columns 2 and 3 of Table 2). $\underline{T}_{memory}$ has an insignificant contribution for the storage task and therefore could be despised without damage for the analysis consistence.

In the storage and recovery processes, a great part of $\underline{T}_{SGPOM}$ is owed to the MO allocation and retention in intermediate buffers inside the SGPOM_DSs (in this case, inside GENERIC). For the developed prototype, no special technique for buffers implementation was used. This could grant an optimised access to the MO contents, entailing better service performance. Extrapolating to the handling of huge MOs, this limitation could become an accentuated bottleneck and, therefore, should deserve a special attention. Another measure source (given, for instance, by $\underline{T}_{buffer}$) could indicate the relative temporal quota associated to the buffering activities, which might be interesting for the media repositories designers.

Table 2. $\underline{T}_{SGPOM}$ (in ms) for the four I/O request types and five MO samples, in SGPOM Basic Configuration, according to (2), (3) and (4).

	Registry	Storage	Recovery	Deletion
Sample1.mov	177.73 (71.9%)	323.5 (33.4%)	148 (29.9%)	116.6 (71.9%)
Sample2.mpg	168.67 (69%)	302.29 (20.8%)	126.12 (14.9%)	111.42 (70.2%)
Sample3.avi	213.36 (75.4%)	317.43 (9.7%)	162.87 (7.7%)	116.54 (65.1%)
Sample4.avi	236 (79.3%)	271.6 (3.69%)	133.2 (2.64%)	110.9 (56.04%)
Sample5.avi	236.69 (79.6%)	387.1 (3.7%)	248.03 (2.8%)	119.32 (54.7%)

4. FINAL REMARKS

This paper reports on some performance validation issues regarding a datastorage multimedia middleware service prototype. Such validation comprises the qualitative analysis of a set of benchmarking results of some established performance metrics obtained during different session cycles.

The performance assessment accomplished was based on temporal measurement values obtained for a service prototype built over a single CORBA product. As the OMG has not yet specified neutral performance evaluation criteria for both existent and future middleware tools, it comes to be very constrained to infer how would be its behaviour over other ORB toolkits. Besides this, there is current a strong appeal towards the deployment of real-time brokering infrastructures [6]. In such context, another set of metrics should be additionally devised in order to embrace other non-flexible temporal requirements.

Plášil et al. [8] emphasise the hardness of investigating benchmarking measures regarding complex actions over a conventional ORB (such as sequences of requisitions), as it implies the need for the separate identification of the implicit influence of certain non-atomic factors. The solution proposed by the authors was to explicitly define elementary actions, such as single marshalling and dispatching operations, which would be of primal relevance to the particular study scenario, such as for throughput, scalability, latency and robustness affairs. For the SGPOM archetype, such a library of basic benchmarking procedures would be very helpful, as for the estimation of the temporal parcels due only to the I/O requests transmission through the framework layers.

Owing to the peculiarities of the proposed service and the lack in the literature of similar published results, the analysis made here does still need to be further improved. In other multimedia branches of research, as the one referring to the human sensorial perception to synchronisation phase-outs between separate media flows, it seems to be more easy to settle down comparison initiatives as they could count on standardised values for certain basic metrics. By other means, it can be stipulated that the performance contemplation introduced here may serve as a first referential for other database benchmarking works, mainly for those regarding the management and access requirements associated to the distributed persistent media data allocated to distributed repository servers.

ACKNOWLEDGEMENT

CNPq has sponsored this research through MSc. and PhD. scholarships to the first author.

REFERENCES

[1] P. Bernstein. Middleware: A Model for Distributed System Services. *Communications of the ACM*, 39(2): 86-98, February 1996.

[2] I. L. M. Ricarte and C. M. Tobar. Towards an Architecture for Distributed Multimedia Databases. *Procs. of ICIIMS*, pp. 65-68, USA, June 1996.

[3] A. L. V. Coelho and I. L. M. Ricarte. A CORBA-based Distributed Multimedia Database Management Layer: Design and Implementation Aspects. *To be published at Procs. of ISCC2000*, France, July 2000.

[4] G. Coulson et al. Supporting the Real-time Requirements of Continuous Media in Open Distributed Processing. *Computer Networks and ISDN Systems*, 27:1231-1246, 1995.

[5] OMG. Persistent Object Service Specification 1.0. TR 97-12-12, December 1997.

[6] D. Schmidt, D. Levine and S. Mungee. The Design and Performance of Real-Time Object Request Brokers. *Computer Communications*, 21(4):294-324, April 1998.

[7] P. Carando. Toward the Development of an Object Request Broker Evaluation Instrument. In *OOPSLA-96 Workshop: Objects in Large, Distributed and Persistent Systems*, 1996.

[8] F. Plášil, P. Tüma and A. Buble. CORBA Benchmarking, TR OMG 98-10-04, Charles University, 1998.

[9] T. Plagemann et al. Adaptive QoS Aware Binding of Persistent Multimedia Objects. In Procs. of DOA, Scotland, September 1999.

[10] IONA Tech. *OrbixWeb Programmer's Guide*, 3.0 edition, November 1997.

Chapter 9

Fault Management in a Multicast Routing Environment

*Kernel Based Tree Protocol, a Case Study**

Mohamed Dâfir ECH-CHERIF EL KETTANI (1)
Younes SOUISSI (2)
(1) Ecole Nationale Supérieure d'Informatique et d'Analyse des Systèmes
ENSIAS – BP 713 – Rabat Agdal – Morocco
e-mail: dafir@ensias.um5souissi.ac.ma
(2) Ecole Mohammedia d'Ingénieurs
Department of Computer Science – EMI – BP 765 – Agdal, Rabat – Morocco
e-mail: souissi@emi.ac.ma

Key words: Inter Domain Multicast Routing, Consensus, Fault Management, Adaptive Communications, Scalability

Abstract: The conception of multicast routing protocols relies generally on the optimisation of Quality of Service parameters. Unfortunately, we rarely find propositions that consider fault tolerance as a criterion to enhance QoS associated to the group: We mean by "fault tolerance" the ability to manage failures of nodes and links of the distribution tree, in the network. A fault tolerant approach at the routing level will allow more reliability at the *network* and *service* levels.

Currently, in a multicast routing environment, when a failure occurs, the solution consists in destroying the distribution tree, and starting from the beginning to construct again the whole distribution tree, which is a costly solution.

We propose a solution avoiding that, by giving fault tolerant extensions to multicast routing protocols. Our proposition is based on the resolution of the *consensus problem*. Consensus problems were studied by Chandra and Toueg: We propose an *inter domain* version of the algorithm. We put the emphasis on scalability issues, due to the hierarchical structure of the network, such as the Internet. We study the case of KBT protocol, a QoS-sensitive protocol, adapted to *inter domain* multicast routing.

* This work is partially supported by PARS MI 16 project

1. INTRODUCTION

In this paper, we are interested in fault management in a multicast routing environment. The solution that we propose gives fault tolerant extensions to multicast routing protocols in general, and particularly to inter domain multicast routing protocols. We mean by "fault tolerance" the ability to manage failures associated to nodes and links that belong to the distribution tree, in the interconnection network. A fault tolerant approach at the routing level will allow more reliability associated to such *network* protocols. It will also associate to the multicast routing protocol a reliable multimedia *service*.

Currently, in a multicast routing environment, when a failure occurs, the solution generally consists in destroying the distribution tree, and starting from the beginning to construct again the whole distribution tree, since every member will join the group again individually, which is a costly issue. This is the case of CBT protocol. Also, resource discovery relies on a mechanism that does not scale [1].

We look for a solution that avoids these problems, and takes advantage of the hierarchical structure of the interconnection network: it supposes that the interconnection network is organised into domains. When the failure of a node occurs, the distribution tree breaks out, but no pruning of the distribution tree happens. We propose a solution that "elects" a new node to replace the failing one, and makes the necessary work to connect all the components of the broken distribution tree to that new node.

Indeed, we first partition the initial distribution tree into sub-trees (one sub-tree per domain). Then each sub-tree will have to decide the value associated to the new "elected node": But, that decided value must be the same among all the domains, which justifies the use of a *consensus* protocol.

Our proposition is based on the resolution of *consensus problem* in an inter domain multicast routing environment. Consensus problems were studied by Chandra and Toueg [3]. We propose an inter domain version of this protocol, solving thus scalability issues, especially in terms of transmitted messages. We study the case of KBT protocol, a QoS-sensitive protocol, adapted to *inter domain* multicast routing. We also analyse the complexity of the solution, and show the advantages of such a policy.

The rest of the paper is structured as follows: In section 2, we present the paradigms associated to KBT protocol, in order to illustrate further fault management in that case. In section 3, we describe the system model, including the failure detector model in an inter domain routing environment. In section 4, we present the consensus algorithm associated to KBT protocol. In section 5, we evaluate the performance of the protocol: We analyse the complexity of the algorithm and compare results with other algorithms. Section 6 corresponds to the conclusion and perspectives of the work.

2. KBT PROTOCOL: STATE OF THE ART

2.1 Overview

Kernel Based Tree protocol is an adaptive inter domain multicast routing protocol. Such a solution has already been presented as part of a hole project, the goal of which was to define an "Inter Domain QoS Multicast Routing Protocol". The first experimentation led to encouraging results [6]. The details of KBT protocol have already been presented in [4]. We present in this paragraph the main characteristics of KBT, in order to treat fault tolerant extensions in next sections.

KBT protocol combines both of the strategies associated to the construction of distribution trees: "Source Based Tree" strategy (noted SBT), and "Shared Tree" strategy (noted ST). These strategies are adapted to two extreme situations of group members configurations referring to the density of group members in the Internet, called "dense" and "sparse" [7]. Combining SBT and ST leads to adaptation properties of the group.

Our strategy consists in grouping initially group members through a core based tree (noted CBT_0, figure 1.a) –according to CBT protocol [1]-, and then defining *"interception points"* on that tree. The position of "*interception points*" may change, according to the evolution of group membership and network characteristics [4]. These points represent the "*kernel*" of the future distribution tree associated to the group. A message sent from a source will transit through 2 trees: the first is a source based tree (SBT) linking that source to *"interception points"*, and the second is the branch of CBT_0, starting at the *"interception point"*. We call the obtained distribution tree **"Kernel Based Tree"** (KBT). Since KBT is an inter domain multicast routing protocol, we suppose that we place one "interception point" per domain involved in the construction of KBT tree. Figure 1.b corresponds to the situation where the (KBT) has 4 "interception points": a source based tree (SBT) links S_1 source to the *"interception points"*.

Figure 1.a. the initial CBT_0

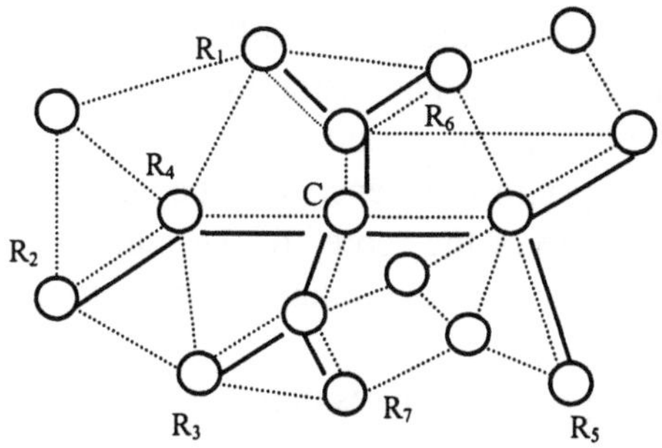

Figure 1.b. the resulting KBT

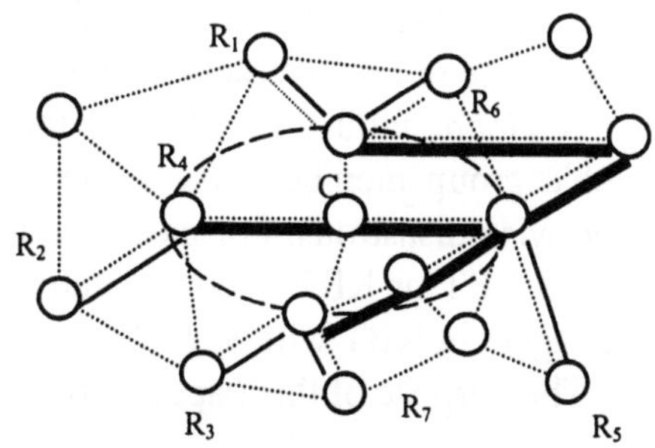

2.2 Management Organisation of KBT Protocol

From a network management viewpoint, and in association to the distribution tree, we can say that KBT architecture is composed of (figure 2):

- The group manager (or group co-ordinator)
- The group initiator (the first node asking for group creation)
- The "core" of CBT_0
- The "Interception Point" (noted SIP for Source Interception Point and RIP for Receiver Interception Point)
- The "Receiver Interception Point" Candidate
- The group member (Receiver or Source)

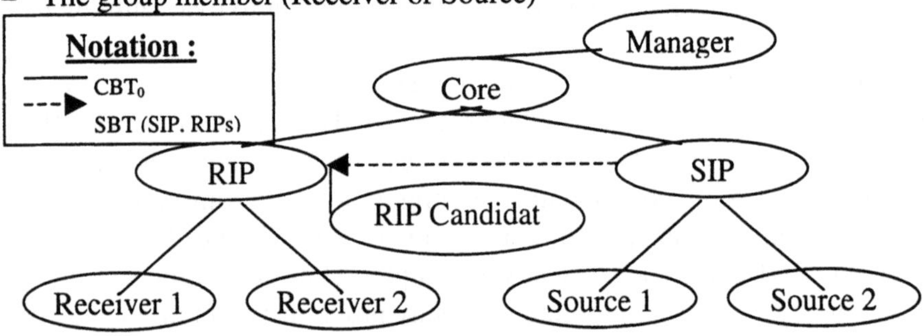

Figure 2. Nodes' dependency

Figure 2 illustrates the dependency graph between KBT nodes. The group manager and the "core" are the main components of the group. The failure of one of them will provoke the failure of the whole group.

3. MODELING FAILURES

3.1 Classification of Failures

We can divide failures into two categories: *local* and *global*.

A *local* failure will not stop the multicast service toward the totality of group members. Its impact on the coherence of the group will be limited. The failure of an "Interception Point" in KBT [5] is an illustration.

A *global* failure is a failure that will stop the multicast service to the whole group members. Its influence on system coherence will be serious, otherwise disastrous. For example, the failure of the "core" of the multicast tree in CBT or KBT is considered as *global*. Also, the failure of the "group manager" in KBT will be *global*.

The purpose of this paper is to deal with *global* failures only.

3.2 Working Rules

The global system works according to 2 main principles:

1. Communications between nodes:

The nodes of the multicast tree associated to a group will be partitioned into domains. Communications inside a domain will be supported by the local multicast tree associated to the group in that domain, whereas we'll have point to point communications at the inter domain level: since a failure is *global*, intra domain communications are not affected by this failure.

2. Detection of a failure:

We associate a failure detector model to our inter domain working environment. Indeed, we suppose that each node of the multicast tree has a failure detector, which is a module that must be able to test the failure of any other node, no matter where the other node is localised (whether inside the same domain, or not). Failure detector model is presented in the next section.

3.3 Failure Detector Model

A failure detector is defined as a module associated to each node of the distribution tree. It outputs the set of nodes of the distribution tree that it currently suspects to have crashed [3]. From a consensus problem viewpoint, two different nodes n_1 and n_2 may not necessarily suspect the same set of failing nodes, because the network is an asynchronous system [10]. So, a failure detector is defined by 2 sets of properties [3]:

Completeness

- **Strong Completeness:** eventually every node that crashes is permanently suspected by *every* correct node.
- **Weak Completeness:** eventually every node that crashes is permanently suspected by *some* correct node.

Accuracy

- **Strong Accuracy:** no node is suspected before it crashes.
- **Weak Accuracy:** some correct node is never suspected.
- **Eventual Strong Accuracy:** there is a time after which correct nodes are not suspected by any correct process.
- **Eventual Weak Accuracy:** there is a time after which some correct process is never suspected by any correct process.

Chandra and Toueg deduce 8 classes of failure detectors:

Accuracy	Strong	Weak	Eventual Strong	Eventual Weak
Strong	P	S	$\Diamond P$	$\Diamond S$
Weak	Q	W	$\Diamond Q$	$\Diamond W$

They also show that in an interconnection network environment (like the Internet), the behaviour of a failure detector associated to a node is characterised by "Strong Completeness" and "Eventual Weak Accuracy" properties: such a failure detector belongs to ◊S class.

However, due to the structure of the interconnection network (organised into domains), we can extend the failure detector model proposed by Chandra and Toueg, based on the class ◊S [3].

Hence, we define a new class of failure detectors, noted by ◊Š, giving hierarchical extension to the failure detector model: This class will act as if each node maintains two failure detectors, the first dedicated to internal (intra domain) communications, and the second for inter domain communications.

This class ◊Š will be characterised by two properties, noted by:

- "*hierarchical strong completeness*": intra domain completeness and inter domain completeness
- "*hierarchical weak accuracy*": intra domain weak accuracy and inter domain weak accuracy

It is important to make this hierarchical distinction, since the interconnection network may be confronted to important heterogeneity characteristics, such as delay, or bandwidth, between intra and inter domain communications.

4. THE CONSENSUS PROTOCOL

4.1 Related Work

Chandra and Toueg proposed solutions to the consensus problem, specific to the different basic classes of failure detectors [3].

We can describe informally a consensus protocol as follows, given a set of correct nodes. At the beginning of the protocol, each correct node is given an input value, and at the end, the non-crashed nodes must have decided on a common output value, belonging to the set of input values.

The following properties define the consensus problem [3]:

- **Termination:** every correct node eventually decides some value.
- **Integrity:** a node decides at most once.
- **Agreement:** no two correct nodes decide different values.
- **Validity:** if a node proposes some value, then this value must have been proposed by some node.

Chandra and Toueg show that the consensus problem can be solved using ◊S class, with at least a *majority* of correct nodes: no blocking happens.

4.2 Overview and Example

We propose an inter domain version of the consensus protocol, solving thus scalability issues. Failure detectors belong to $\Diamond\check{S}$ class.

For example, suppose that we have a group working under KBT protocol (figure 3). We consider that "core" failure is a *global* failure. If the failure of the node C happens (C is the "core" of the group), we'll first partition the initial multicast tree CBT_0 into sub-trees, according to domains (figure 3).

At the end of this first step, each sub-tree proposes one potential new "core" candidate to the group. The choice of this new "core" candidate is operated thanks to QoS considerations, locally to each domain.

However, the final value of the new "core" must be the same among all the sub-trees: a consensus protocol, at the inter domain level, will let all the sub-trees decide on a common final value.

Figure 3. KBT multicast tree (4 domains, 1 source, 7 receivers)

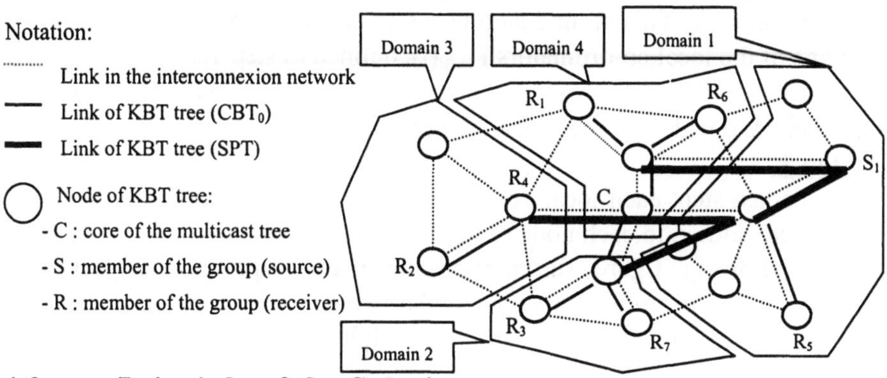

4.3 Principle of the Solution

The consensus problem -based on $\Diamond S$ class- can be solved with the hypothesis that the maximum number of failing nodes is less than half of implied nodes. Chandra and Toueg show that, at any time, all the nodes may be erroneously added to the list of suspects. However, there is a correct node and a time after which that node is not suspected to have crashed [3]. In the case of $\Diamond\check{S}$ class, we make the same suppositions, since:

- A node of $\Diamond\check{S}$ class has at the same time two failure detectors.
- The properties of each failure detector are taken from $\Diamond S$ class.

So, the proposed consensus protocol can be solved with the hypothesis:

- Each implied *correct* domain proposes a local value. So, it is supposed not to be a *failing* domain. In a *failing domain*, the maximum number of failing nodes is less than half of the nodes of the group in that domain.
- A majority of implied domains in the consensus process are not failing: they are correct.

4.4 Description of the protocol

The proposed protocol uses the rotating co-ordinator paradigm [2] [9], at the *intra* and *inter* domain levels. Our solution is summarised in 3 main steps (figure 4). Details are presented in section 4.5.

Step 1:

We start with a local consensus in each domain D_i of the interconnection network. That's why each domain D_i will have a local co-ordinator. The rotation of local co-ordinators will act in an asynchronous way in terms of rounds. Each round is characterised by the four classical steps of consensus protocol [3]. At the end of this first step, each domain D_i proposes a common value -noted Δ_i- to all the nodes of that domain among the proposed values $\Delta_{i,j}$. The local co-ordinator of domain D_i is noted $P_{local}(i)$.

Step 2:

We apply the consensus algorithm at the inter domain level, on the set of local co-ordinators $P_{local}(i)$ of domains D_i. Hence, each domain D_i proposes an initial value Δ_i. This second step needs a global co-ordinator, which will meet one of the local co-ordinators $P_{local}(i)$, defined in step 1.

At the end of this step, each local co-ordinator has received the common final value Δ, decided by the global co-ordinator.

Step 3:

In this last step, each local co-ordinator $P_{local}(i)$ in domain D_i broadcasts the decided value Δ to each node of his domain.

The following figure summarises the main steps of the proposed protocol. In this example, we suppose that we have three domains D_1, D_2, and D_3. We also suppose that $P_{local}(1)$ is the global co-ordinator.

Figure 4. Illustration of inter domain consensus protocol (Steps and Phases)

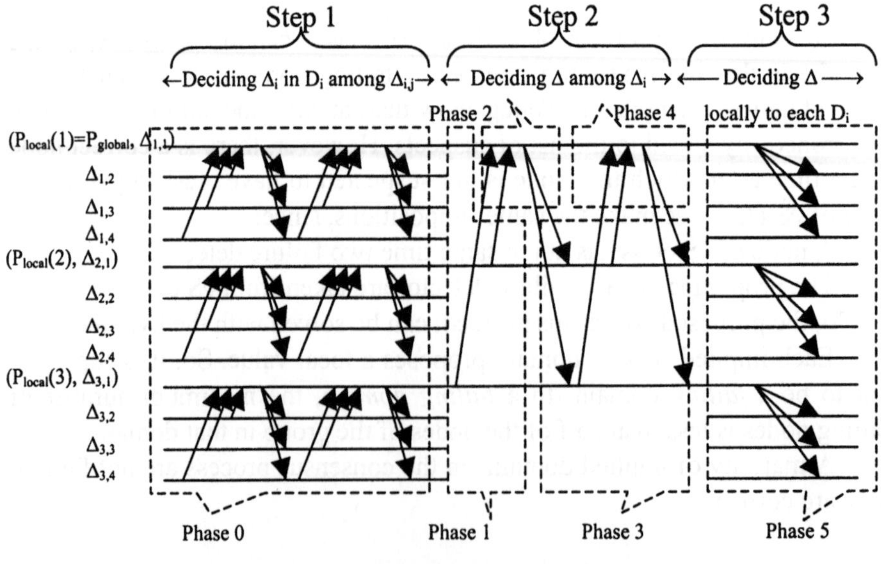

4.5 Algorithm

Each local co-ordinator node $P_{local}(i)$ executes the following algorithm. It will either execute that algorithm totally or partially, depending on the role of that node in the consensus protocol, which will be one of the following:

Local co-ordinating node, noted $P_{local}(i)$, belonging to a domain D_i

Global inter-domain co-ordinating node, noted P_{global}.

Notation:

Local_consensus(): consensus algorithm of Chandra and Toueg, executed in the domain mentioned as a parameter. It returns the estimated value Δ_i.

F_Broadcast_local(): function that broadcasts the estimated value $\Delta = Estimate_{global,Plocal(j)}$ from a local co-ordinator to each node of its domain.

F_Broadcast_global(): function that broadcasts the estimated value Δ from the global co-ordinator to local co-ordinator of each domain.

F_rcv_msg_bcast(): reliable function launched by each domain's local co-ordinator, waiting for the estimated value Δ from the global co-ordinator. It broadcasts reliably (locally to each domain) the decided value.

Decide(): function that makes the final decision.

$Estimate_{global,Plocal(i)}$: the estimate of the global expected value Δ, in $P_{local}(i)$ node of domain D_i.

Figure 4 summarises the main phases of the algorithm. The algorithm of the "inter_propose" procedure of the inter domain consensus protocol is:

Procedure inter_propose($P_{local}(i)$, Δ_i)

Phase 0 (executed by each local co-ordinator)

$Estimate_{global,Plocal(i)}$ = Local_Consensus(D_i) = Δ_i; $State_i = 0$; $R = 0$; $TS_i = 0$;
while ($State_i = 0$) do
{R= R+1; P_{global} = (R mod G)+1;}

Phase 1

Send ($P_{local}(i)$, R, $Estimate_{global,Plocal(i)}$, TS_i) to P_{global};

Phase 2

if ($P_{local}(i)$= P_{global}) then
wait until [for $\lceil (G+1)/2 \rceil$ nodes$P_{local}(j)$, with $j \neq i$:
receive ($P_{local}(j)$, R, $Estimate_{global,Plocal(j)}$, TS_j)];
$msgs_i[R_i]$ = {($P_{local}(j)$, R, $Estimate_{global,Plocal(j)}$, TS_j) /
$P_{local}(i)$ received ($P_{local}(j)$, R, $Estimate_{global,Plocal(j)}$, TS_j)};
t = max{TS_j};
$Estimate_{global,Plocal(i)}$ = $Estimate_{global,Plocal(j)}$ / {t=TS_j and
($P_{local}(j)$, R, $Estimate_{global,Plocal(j)}$, TS_j)∈ $msgs_i[R_i]$};
send ($P_{local}(i)$, R, $Estimate_{global,Plocal(i)}$) to all $P_{local}(j)$, with $j \neq i$;

Phase 3

wait until{(received(P_{global}, R, $Estimate_{global,C}$)) or suspect failure of (P_{global})};
if (received(P_{global}, R, $Estimate_{global,C}$)=true) then
$Estimate_{global,Plocal(i)}$=$Estimate_{global,Pglobal}$; TS_i=r;send(P_{local}(i),R,ack) to P_{global};
else
send(P_{local}(i), R, nack) to P_{global};

Phase 4

if (P_{local}(i)= P_{global}) then
wait until (for $\lceil (G+1)/2 \rceil$ nodes P_{local}(j), with $j \neq i$:
received(P_{local}(j), R, ack) or received(P_{local}(j), R, nack));
if (for $\lceil (G+1)/2 \rceil$ nodes P_{local}(j), with $j \neq i$: received(P_{local}(j),R,ack)=true) then
F_Broadcast_global(P_{local}(i), R, $Estimate_{global,Plocal(i)}$,decide);

Phase 5 (executed by each local co-ordinator)

when (F_rcv_msg_bcast (P_{local}(i), R, $Estimate_{global,Plocal(j)}$,decide))
if ($State_i$ = 0) then
Δ = $Estimate_{global,Plocal(j)}$;
Decide(Δ); $State_i$ = 1; F_Broadcast_local(P_{local}(i), R, Δ,decide);

4.6 Proof of Correctness

We proof that the proposed algorithm verifies the four properties that characterise the consensus problem (Section 4.1). The adopted approach to proof these properties is similar to the one chosen by Chandra and Toueg. For more information, a detailed description exists in [3]. A failure detector of class ◊Š, is characterised by intra and inter domain strong completeness, and intra and inter domain eventual weak accuracy

4.6.1 Property 1

Termination: every correct node eventually decides some value.

Proof: It means that no node is blocked on a **wait** statement.

Let us suppose that a node blocks on a wait statement, and let us take the earliest round and point on that round where this blocking occurs.

If it is local consensus, it can't be blocking, by consensus [3].

If it is a waiting statement in phase 2, 3, or 4 of the algorithm, the "hierarchical strong completeness" property will ensure that the waiting will stop. In the case of a waiting statement of the global co-ordinator, then the hypothesis of a majority of correct domains (see section 4.1) guarantees the reception of at least a majority of messages. So, no blocking occurs.

4.6.2 Property 2

Integrity: a node decides at most once.

Proof:

The algorithm satisfies the integrity property, because each node decides at most once (*Phase 5*): the *decide* function is called once.

4.6.3 Property 3

Agreement: no two correct nodes decide different values.

Proof:

If no process decides, the property is trivially true.

If any process decides, let us take the time at which the first decision has been taken. The global inter domain co-ordinating node has eventually received a majority of acknowledgements. This implies that $\lceil (G+1)/2 \rceil$ of local co-ordinating nodes detain this estimate. We proof that this estimate holds indefinitely. We apply the demonstration in [3], considering in that case each domain as a process. The proof is by induction on the round number. More details can be found in lemma 19 of [3].

In phase 5, since each correct local co-ordinating node has previously executed the *F_rcv_msg_bcast* function, it will broadcast this value to all the nodes of the domain. The decided value must have been previously decided by the global co-ordinator.

4.6.4 Property 4

Validity: if a node proposes some value, then this value must have been proposed by some node.

Proof:

From the algorithm, the decided values are taken from estimates at two levels:

Intra domain level: all the estimates $\Delta_{i,j}$ received by any local co-ordinator in phase 0 are proposed values from the nodes of the same domain. The result is a proposed value Δ_i at the intra domain level.

Inter domain level: all the estimates that a global co-ordinator receives in phase 2 are proposed values from local co-ordinators. So, the value decided by a global co-ordinator must be the value proposed by some node.

As a consequence, the algorithm proposed in section 4.5 solves Consensus using ◊Š class.

5. PERFORMANCE EVALUATION

5.1 Complexity Analysis of the Solution

Suppose that we have G domains, and 'n' nodes in each domain. The complexity of the algorithm proposed in section 4.5 in terms of exchanged messages is in $O(G^2)+O(Gn^2)$:

Indeed, the local consensus operation (phase 0) consumes $4(n-1)+(n-1)^2$ messages per domain: $4(n-1)$ is the number of exchanged messages between the nodes and the local co-ordinator, in order to take the local consensus value. $(n-1)^2$ is the number of exchanged messages due to the final local broadcast function [3]. This broadcast function is supposed to be reliable. So, the global number of messages is equal to $G*[4(n-1)+(n-1)^2]$.

The global inter domain consensus operation (phases 1, 2, 3, and 4) consumes $4(G-1)+(G-1)^2$. The reasons are the same that in local consensus problem, considering in that case a domain instead of a node.

The final broadcast function in phase 5 consumes $(n-1)^2$ messages per domain: the broadcast function is a reliable broadcast function. So, the total number of messages is $G*(n-1)^2$.

Hence, the complexity of the whole algorithm is in $O(G^2)+O(Gn^2)$.

5.2 Comparative Analysis

There is an improvement, concerning scalability, in terms of exchanged messages in comparison with Chandra and Toueg algorithm, which is in $O(G^2n^2)$.

We can also find that our solution is better than the one proposed by [8], where the complexity is in $O(G^2+n^2+Gn^2)$.

6. CONCLUSION

Fault-tolerant management in a *multicast routing environment* is a new issue in *inter domain* multicast routing.

In this paper, we have studied the case of KBT protocol, which is a generic adaptive inter domain multicast routing protocol. After this, we presented the main concepts of the protocol, in terms of construction of the distribution tree, and distributed management capabilities: we showed that KBT protocol can integrate at the routing level a certain level of control and management. This offers a reliable multimedia service associated to the multicast group.

Then, we associated a failure model to the system, by presenting a classification of failures associated to the system, and then proposing a new class of failure detectors, adapted to the hierarchical structure of the interconnection network. We finally showed why the problem can be considered as a classical consensus problem.

We also showed how we can optimise that classical consensus problem, through a hierarchical approach. We proposed the algorithm associated to the hierarchical consensus problem, and proofed its correctness.

Performance evaluation of the solution included complexity analysis, and comparison with the consensus protocol: we obtained good results.

The applications of our solution can be extended to protocols others than KBT: "core" election mechanism in CBT and PIM-SM for example. Other applications can be seen in any hierarchical inter domain distributed environment that needs to maintain a global state, or to elect a common value between components of the system.

REFERENCE

[1] A. Ballardie, "Core Based Trees (CBT ver. 2) Multicast Routing Architecture", Internet RFC 2201, September 1997.

[2] T.D. Chandra; and S. Toueg, "Time and Message Efficient Reliable Broadcasts", In Proceedings of 4th International Workshop on Distributed Algorithms (WDAG-4), Springer Verlag, November 1996.

[3] T.D. Chandra; and S. Toueg, "Unreliable Failure Detectors for reliable distributed systems", Journal of the ACM, vol 43, n°2, pp-225-267, 1996.

[4] M.D. Ech-Cherif El Kettani, "KBT Protocol Architecture", Technical Report n° 02/98, Ecole Mohammedia d'Ingénieurs, Rabat, Morocco.

[5] M.D. Ech-Cherif El Kettani; and Y. Souissi, "An Adaptive Architecture for Inter Domain QoS Multicast Routing", in French, in Calculateurs Parallèles: Routage dans les réseaux, vol 11, n°1/99, Hermes, pp-59-86.

[6] M.D. Ech-Cherif El Kettani; and Y. Souissi, "Performance Evaluation of Multicast Routing Protocols: Comparative Study and Proposition of an Adaptive Solution", CFIP'99 Proceedings, Nancy, France, 26-29 April 1999, Hermes, pp-317-332.

[7] C. Huitema, "Routing in the Internet", Prentice-Hall, Englewood Cliffs, N.J., 1995.

[8] F. Nguilla Kooh, "Hierarchical Approach for Solving Agreement Problems in Wide Distributed Systems", In Proc. of Second IASTED International Conference on Parallel and Distributed Computing and Networks, December 14-16 1998 Brisbane, Australia.

[9] R. Reishuck "A New Solution for the Byzantine General's Problem", RJ 3673, IBM Research Lab., Nov. 1982.

[10] M. Raynal, "Synchronisation and Global State in Distributed Systems", in French, Eyrolles Editions, 1992.

Chapter 10

Traffic Management in Isochronets Networks

Kelvin Lopes Dias[1], José Augusto Suruagy Monteiro[2] and Danilo Florissi[3]

1Universidade Federal de Pernambuco-UFPE,Centro de Informática
P.O. Box 7851, CEP: 50732-930, Recife,Pernambuco – Brazil
E-mail: kld@cin.ufpe.br
2 Universidade Salvador – UNIFACS, Engineering Departament
E-mail: suruagy@unifacs.br
3 Columbia University, Departament of Computer Science
Distributed Computing and Communications Lab
450 Computer Science Bldg, New York, NY, USA 10027
E-mail: df@cs.columbia.edu

Key words: Quality of Service, Isochronets, RDMA, Leaky Bucket, Virtual Clock

Abstract: This paper presents an evaluation of a traffic management mechanism for high speed networks called RDMA (Route Division Multiple Access), developed as part of the Isochronets, a novel architecture for high speed networks. RDMA enables routes to designated destinations during time intervals (green-bands) available periodically. We evaluated the support RDMA offers to applications quality of service (QoS) represented by delay and jitter and its network congestion prevention effectiveness. RDMA is compared to two other well-known mechanisms: Virtual Clock and Leaky Bucket. RDMA showed itself to be satisfactory as an access control mechanism for applications with strict QoS requirements. In RDMA, memory resource needs at network elements is minimized; traffic sources or network peripheral elements are responsible for the use of a large quantity of memory.

1. INTRODUCTION

The multiservice structure of the broadband networks will carry a mix of traffic types ranging from traditional data communications to high quality video transmission, like: telemedicine, video on demand, and virtual reality.

Fundamental for the success of this service integration is the concept of traffic management, which comprises a set of functions which aim at satisfying Quality of Service (QoS) requirements for applications, especially, during periods of network congestion.

Access control mechanisms, part of the traffic management framework, prevent high speed network congestion by monitoring and enforcing traffic contract established during call admission. Basically, these mechanisms can act in two ways: policing and shaping the traffic. In the first case, a policer can either drop packets violating contracts or mark them as low priority which may lead to discarding in congested nodes. In the second case, the mechanism can delay the packets at the sources to comply with the contract. Discarding can occur due to buffer scarcity [11]. The access control can be implemented at the traffic sources [8] and at network switches [10]. Besides the essential function performed by these mechanisms, they can degrade the QoS, for example, increasing end-to-end delay when acting as shaper, or discarding packets when acting as a policer.

In this paper, we analyse a new access control mechanism called RDMA (Route Division Multiple Access) proposed as part of a new architecture for high speed networks, Isochronets[3], from the view point of shaping. Our objective is to evaluate the support RDMA offers to application performance requirements and its network congestion prevention effectiveness. The results obtained for RDMA are compared to two other access control mechanisms: Leaky Bucket [12] and Virtual Clock [13]. This paper is structured as follows: section 2 describes Isochronets/RDMA, as well as, Leaky Bucket and Virtual Clock. In section 3 performance evaluation results of these mechanisms are presented. Finally, section 4 presents the conclusions and proposals for future works.

2. MECHANISMS

2.1 Isochronets and Route Division Multiple Access (RDMA)

Isochronets are a novel switching architecture that coordinates traffic motions in the network over time, similar to how green waves coordinate traffic motion in roads. In the latter scenario, coordination is accomplished through synchronized traffic lights to create routes for uninterrupted and collision-free motions from source to destinations. In Isochronets, coordination is accomplished by enabling a set of routes by means of periodic configuration of network switches. Switches set up routing trees, that is, routes from all nodes to a given destination. A frame is switched by following its path through an enabled tree.

Figure 1 ilustrates the idea. During the green band (shaded), a frame transmitted by a source will propagate down the routing tree to the destination root. If no other traffic contends for the tree, it will move uninterrupted, as depicted by the straight line.

Multiple simultaneous routing trees can schedule transmission in parallel (have simultaneous green bands), depending on the network topology. For an extreme example consider a fully connected network: all trees to all nodes

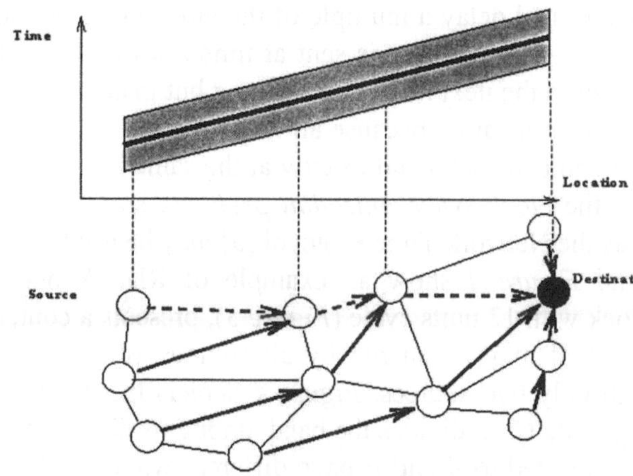

Figure 1. Motion Via Green Band.

can be simultaneously active without interference. In more realistic examples, significant parallelism can be accomplished. *Figure 2* shows two non-interfering routing trees.

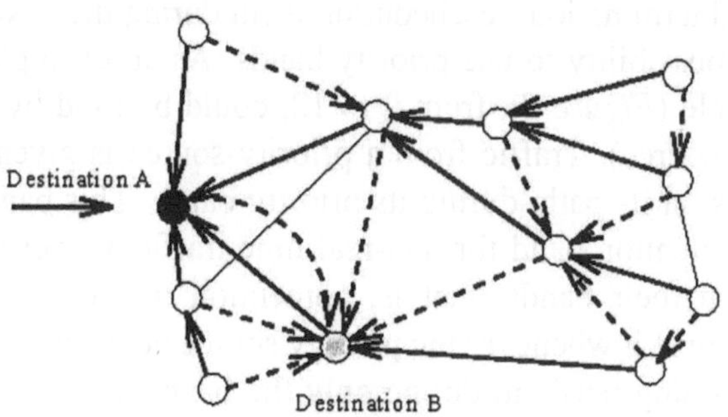

Figure 2. Multiple Non-interfering Trees.

The Isochronets coordination mechanism is called *Route Division Multiple Access* (RDMA) because it divides network bandwith among routing trees. RDMA is employed in Isochronets to assign bands (or bandwidth slices) to routing trees, that is, network switches are configured periodically to build different routing trees. Each configured tree lasts a time band (green band) after which the next time band and corresponding tree replaces it. A sequence of bands building trees that cover each destination once and only once form a *cycle*, which is repeated periodically.

The green band is maintened by switching nodes through timers synchronized to reflect latency along tree links. This problem of band synchronization is solved by manipulating the link propagation delays in the network. There is a device in each link that can be set to increase the link delay. The protocol compute the real link delay and then increases it using the device to make final delay a multiple of the clock cycle period. Because of this manipulation, if something is sent at time *t* within the cycle from a given node, it reaches the next node still at time *t* but many cycles later. This fact trivializes synchronization because all that is necessary after link delays are set is to start and end each band exactly at the same moment within each cycle. To solve the *clock synchronization problem*, any of the traditional protocols such as the Network Time Protocol [5] may be used.

Figure 3 and *Figure 4* show an example of RDMA operation. The hypothetical clock with 12 units cycle (*Figure 3*), presents a contention band allocation (band that may be shared by all sources simultaneously) for a routing tree with only two sources. *Figure 4* depicts the band allocation at the intermediary node *C*. *C* divides the band among its descendents *A* and *B*. The band sizes allocated to *A* and *B* have different widths and only overlap partially. When contention occurs, the collision resolution mode used is designated in terms of the signs: –, +, and ++. In RDMA–, only one of the colliding frames proceeds, while the others are discarded. In RDMA+, one colliding frame proceeds while the others are buffered, but only up to the band duration. RDMA++ operates similarly to RDMA+, but also stores frames beyond band termination, rescheduling them during the next band.

There is also a possibility to use priority bands. As an example, the last quarter of clock cycle (*Figure 3*), from 9 to 12, could be used by a priority source (e.g., video source). Traffic from a priority-source is given the right of way, by switches on its path, during its priority band. This band will not overlap with the contention band for non-real time traffic, however, priority sources do not own their bands, that is, contention traffic may access a priority band and utilize it whenever the priority source does not.

RDMA dimensioning needs to define only the band size for the flow to be controlled. This choice takes into account the mean transmission rate required by the source. The band size may represent a percentage of RDMA basic cycle, defined as 125μs [3]. The flow band size will have to take into account the output link capacity, that is, the proportion of the cycle allocated will define the proportion of output link capacity allocated. For example, if a link has 150Mb/s capacity and a source requires a mean rate of 75Mb/s, the allocated band for this source may be 75/150 = 50% of the 125μs RDMA cycle, that is, the band size will be 62,5μs.

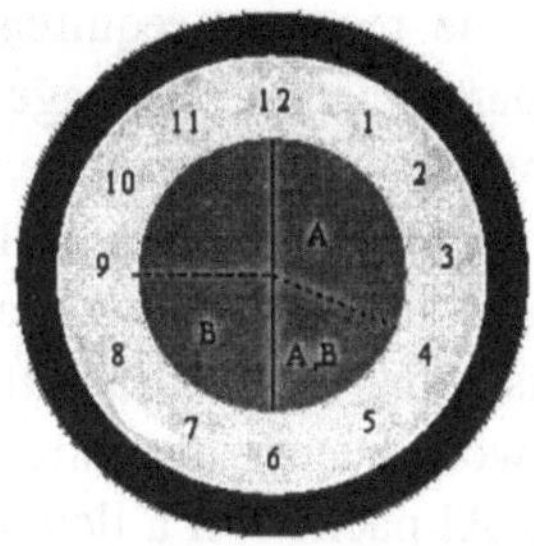

Figure 3. RDMA Allocation Example.

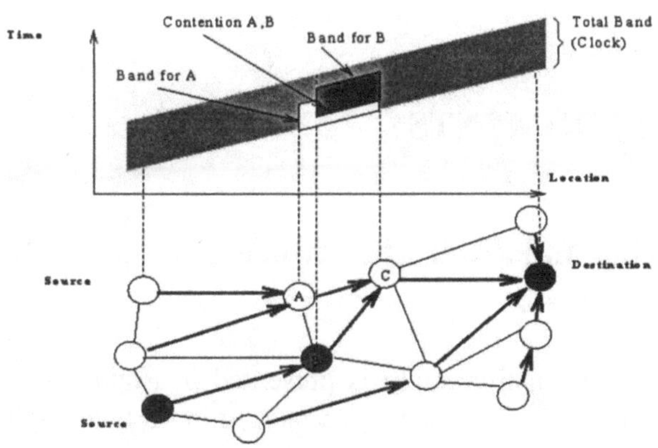

Figure 4. RDMA: Contention Band.

2.2 Leaky Bucket

This mechanism was proposed initially in [12] and it is used as the access control mechanism for ATM networks. Leaky Bucket consists of a token/credit generator, a pool of tokens/credits, and a buffer (this buffer can be optional). Credits are generated every λ seconds and stored in the pool of maximum size B credits. An arriving cell (packet with fixed size used in ATM networks) uses one credit from the pool and leaves the system immediately. An arriving cell that finds the credit pool empty joins the queue if the buffer is not full. When the queue is not empty (which means that the credit pool is empty) and a credit is generated, one cell takes the credit and leaves immediately.

2.3 Virtual Clock

Virtual Clock mechanism was designed as part of network architecture called Flow Network[13]. The resource requirements for each source are expressed in terms of two parameters: an average rate (AR) and an average interval (AI) to check the flow conformance. The network access is controlled by a mark called *virtualclock* given to the packets. Each data flow has a *virtualclock* which ticks at every packet that may access the network for that flow; the tick step is equal to the mean inter-packet gap (assuming a constant packet size). Network access is controlled in the following way: upon transmiting each AR x AI packets, if a flow sends packets according to its specified average rate, its *virtualclock* reading should be in the vicinity of the real time. Otherwise, the flow will be penalized by delaying or discarding its packets.

3. EXPERIMENTS

3.1 Delay and Jitter at Network Access for Real Time Traffic

The model used in this study is presented in *Figure 5*. An individual source generates ATM cells with data packed in 48 byte payloads. The link capacity to access the network is 150Mb/s. The following metrics were considered: access delay, defined as the difference between the time when the cell enters the network (source transmits cell) and the time when the cell was generated; jitter, defined as the difference between the network access delay of two consecutive cells; and, finally, the mean occupation in the infinite buffer [1] of *Figure 5*.

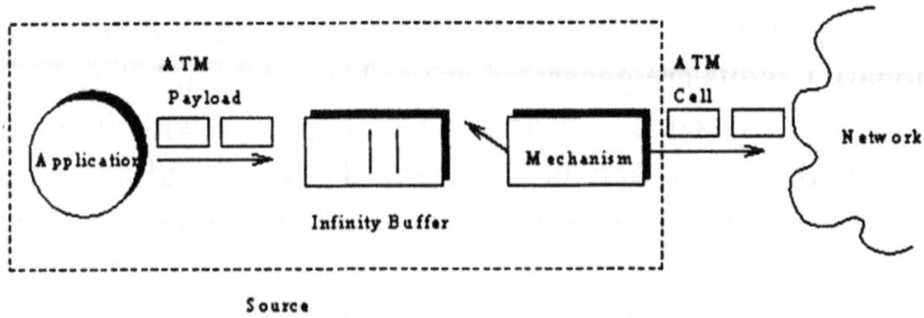

Figure 5. Source Model.

[1] The term "infinite buffer" means that the buffer space is sufficient to avoid losses due to the shaping effect caused by each one of the three simulated mechanisms.

3.1.1 Voice Source

For the voice source we utilized the ON-OFF model according to the conclusions in [2], with the following parameters: peak rate=167 cells/s, mean rate = 60 cells/s, burstiness(peak rate/mean rate) =2.78, and mean burst length = 60 cells. In RDMA dimensioning, the band size for this traffic was set to $1.9\text{x}10^{-2}\mu s$ for a corresponding 125μs RDMA cycle. Leaky Bucket used a credit pool equal to 60 (mean burst length) and the credit rate corresponding to the voice mean transmission rate of 60 cells/s. For Virtual Clock the mean rate, 60 cells/s, was its rate parameter (AR) and the time between checks corresponded to the source average transmission time (ON state): 360ms or β^{-1} [2]. The parameters have been set to provide a fair comparison of these mechanisms.

Table 1. Voice Source

Mechanism	Mean Delay	Max Jitter	Mean Jitter	MeanOccupation
RDMA	0.0001s	0.000116 s	0.000009 s	0.5 cell
L. Bucket C=1.2	1.0058 s	0.0138 s	0.00523 s	65.57 cells
L. Bucket C=1.4	0.2930 s	0.0119 s	0.00242 s	22.29 cells
V. Clock	0.2814 s	0.6501 s	0.0335 s	27.22 cells
V. Clock C=1.2	0.1930 s	0.6501 s	0.0144 s	19.03 cells

As done in [9], an overdimensioning factor C > 1 is added to the Leaky Bucket mechanism. Under this scheme, the Leaky Bucket credit rate is equal to C times the negotiated mean traffic rate. We used the same idea for Virtual Clock rate parameter, AR. Simulation results are showed on *Table 1* and they are accurate within 6% with a 95% confidence level using batch means technique.

RDMA has the smallest delay while Leaky Bucket has the largest. RDMA delay is exclusively due to waiting time for the band to the destination. The maximal delay suffered by RDMA will be the inactive band duration (cycle duration minus band duration), queueing of previous cells waiting for the same band, and the cell transmission time. It is important to emphasize that only cells to the same destination are queued waiting for a band. Leaky Bucket delay is essentially caused by absence of tokens in the credit pool (a similar result was obtained in [7] when it was verified a serious delay due to the Leaky Bucket shaping process at an user end station).

RDMA can satisfy the requirement of end-to-end delay, typically varying from 0 to 150 ms [6]. It is important to say that both Leaky Bucket and Virtual Clock could have a smaller delay than RDMA. However, their parameters must be overdimensioned, which will cause congestion inside the network and significantly degrade the QoS of other connections. We have choosen to dimension all mechanisms based on the average rate promised by the source. Jitter and occupation for RDMA was smaller as well. For RDMA, the biggest jitter occurs when the second cell arrives right when the band closes and is forced to wait until the next active band to enter the network. RDMA fixes the cycle size at 125μs to satisfy the sampling rate of voice sources. In the voice experiments, the cell inter arrival time is 1/60 cells per second=16ms and the periodicity of active band is aproximately 125μs. RDMA services at least one voice samples per cycle.

3.1.2 Video Source

We assume that video traffic is produced by encoding subsequent video frames at a rate of 30 frames/s. The amount of information (number of bits) in a frame is generated according to a first order autoregressive model as proposed in [4], representing a videophone application, with mean rate equal to 3.9Mb/s (0.52 bits/pixel) and a peak rate equal to 10.68 Mb/s (1.41 bits/pixel). With this model, the number of bits/pixel for frame *n*, denoted by $\lambda(n)$, is given as follows: $\lambda(n) = a\lambda(n-1) + b\omega(\eta)$, where *a* and *b* are constants and $\omega(\eta)$ is a white Gaussian noise component with mean η and variance 1. Here we use the following values: $a \approx 0.8781$; $b \approx 0.1108$ and $\eta \approx 0.572$. The number of pixel/frame is assumed to be 250,000 and the expected value $E(\lambda)$, of $\lambda(n)$ is 0.52 bits/pixel.

RDMA band size needs to accomodate a mean rate equal to 3.9Mbps/(8bits×48 bytes) = 10,156 cells/s. For a 150Mb/s link capacity, the band occupies 2.6% of the basic RDMA cycle, that is, 3.25μs from a 125μs cycle. Leaky Bucket had credit rate equal to 10,156 cells/s and the credit pool size configured to the mean and maximal burst size, that is, (0.52×250,000)/(8×48) = 339 and (1.41×250,000)/(8×48) = 918 cells, respectively. Virtual Clock used the video mean rate as the AR parameter and a intercheck time equal to 33ms (video frame periodicity).

Table 2 shows the results obtained for this experiment, which are accurate within 8% for a 95% confidence level using batch means. The delay suffered by RDMA is only greater than the one by Leaky Bucket for *C=1.4* and the one by Virtual Clock for *C=1.2*. The maximal jitter introduced by the mechanisms is related to the parameter used to dimension them. For example, the maximal jitter for Leaky Bucket corresponds to the inter-credit arrival time, while for Virtual Clock it is due to the maximal penalty delay that a video cell may suffer. In RDMA, again, jitter value is due to the inactive band size.

Mean occupation was relatively high for all mechanisms, but mainly for Leaky Bucket with C = 1.15. This value was substantially bigger than the mean video frame (339 cells). Virtual Clock had the smallest occupation because we imposed a limit in the maximal time that a cell could be delayed, that is, 33 ms (video frame periodicity). RDMA had a satisfactory occupation compared to the others, even using only a conservative dimensioning, that is, sizing the band size for the promised video mean rate.

Table 2. Video Source

Mechanism	Mean Delay	Max Jitter	Mean Jitter	MeanOccupation
RDMA	0.00787 s	0.0001217 s	0.000041 s	125.65 cells
L. Bucket C=1.15	0.16115 s	0.000086 s	0.000010 s	1,282 cells
L. Bucket C=1.15 Credits=918	0.13198 s	0.000086 s	0.000008 s	947.34 cells
L. Bucket C=1.4	0.0166 s	0.00007 s	0.000002 s	210.79 cells
V. Clock	0.00904 s	0.03300 s	0.000102 s	88.22 cells
V. Clock C=1.2	0.00456 s	0.03300 s	0.000027 s	45.19 cells

3.2 Congestion Control and Delay for MMPP Traffic

The primary aim of this study is to understand RDMA congestion prevention effectiveness. The congestion degree is indicated *qualitatively* by means of number of cell losses and memory requirements for buffers present at the network internal switches. The accesss control mechanisms are used at peripheral nodes of the network topology proposed for the simulations. Besides that, we investigated the delay caused by the mechanisms at entrance of the network due to the shaping process.

The network topology used in this study is formed by local area networks and wide area networks, interconnected by a high-speed backbone. The corresponding access control mechanisms (RDMA, Leaky Bucket and Virtual Clock) are implemented inside the peripheral nodes. The resources, such as: bandwidth and available memory in switching elements are the same for simulations of all mechanisms.

In order to model the aggregated traffic generated by each one of the peripheral networks, we used a two-state Markov Modulated Poisson Process (MMPP) [1]. Each ON-OFF source that belongs to a MMPP has the following characteristics: peak rate = 10Mb/s, mean rate = 1Mb/s, burstiness= 10, mean burst length = 100 cells, and cell length = 424 bits. The MMPP is a reasonable model for this experiment because the sources generate periods of overload, even within the traffic contract.

The RDMA dimensioning uses a band size corresponding to the average rate specified by the aggregate traffic. The credit rate of Leaky Bucket is configurated with the mean rate of the aggregated traffic multiplied by the factor $C = 1.4$, and the credit pool size corresponds to the MMPP burst mean length. Virtual Clock uses the mean aggregate rate as its rate parameter and

the period between checks are presented by graphs. We limit the maximum period that the flow should wait when penalized by Virtual Clock to 0.1 ms.

3.2.1 Results for Internal Memory Utilization

For this simulation we utilize 10 batches each one with 6.8×10^7 cells. In our experiments, the flow could travel in two ways along the backbone: one routing tree or two routing trees, according to Isochronets philosophy. In the first case, the sources LAN_1 and LAN_2, transmit data to the destination LAN_4. Both flows contend for the same output link at C_3 node, which leads to the

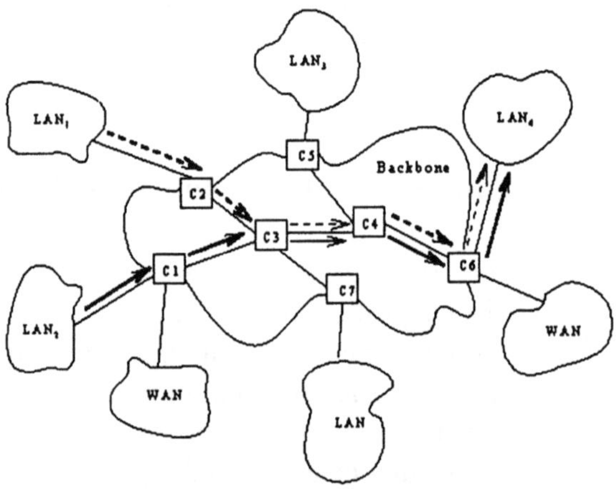

Figure 6. One Routing Tree.

C_4 node and, finally, to the corresponding peripheral node belonging to the backbone that do the direct connection with destination network (*Figure 6*). According to the Isochronets philosophy this corresponds to a routing tree. In the second configuration, the bands at LAN_1 and LAN_2 do not overlap in time and the destinations are LAN_3 and LAN_4, respectively. In this case we have two routing trees (*Figure 7*)

The curves obtained with the experiments consider two load regions for the traffic accessing the backbone: an underload region with the traffic within the contract for the mean rate specified by source, and an overload region where the source generates traffic above the contract. The mechanisms are configured to a load of 50% of link capacity and such configuration remains unchanged for all loads used to plot the curves. We used a backbone with 150Mb/s internal link capacity.

It is important to say that in the simulated model, RDMA is also at internal nodes according to the Isochronets operation. The models for Leaky Bucket and Virtual Clock do not limit the bandwidth inside the backbone, because these two mechanisms are present only at backbone periphery. In this way, all capacity of output links at internal nodes is available to simulate Leaky Bucket and Virtual Clock.

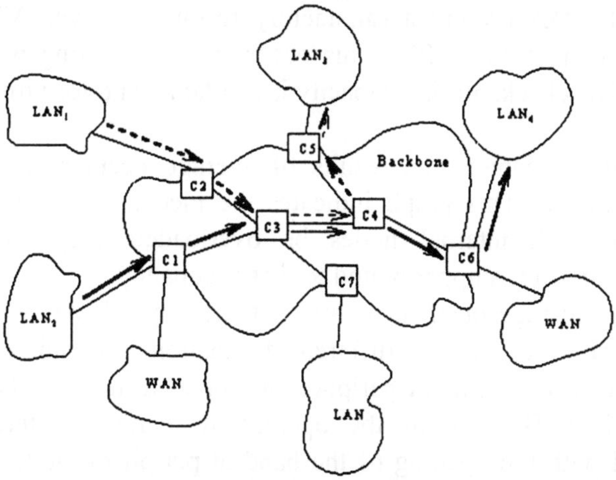

Figure 7. Two Routing Trees.

The internal switching elements are configured with a buffer of 20,000 cells. This value is sufficient to avoid losses at backbone peripheral nodes for loads bellow those used to configure the mechanisms parameters.

The *Figure 8* depicts the losses suffered by each mechanism for one of the peripheral nodes (C_2) that receive the flows. The losses occur after an offering load of 50% (used to configure the mechanisms). The average number of rejected cells is very high for RDMA and Leaky Bucket because this is an instability region, where the service rate of mechanisms is less than the arriving rate of flow cells.

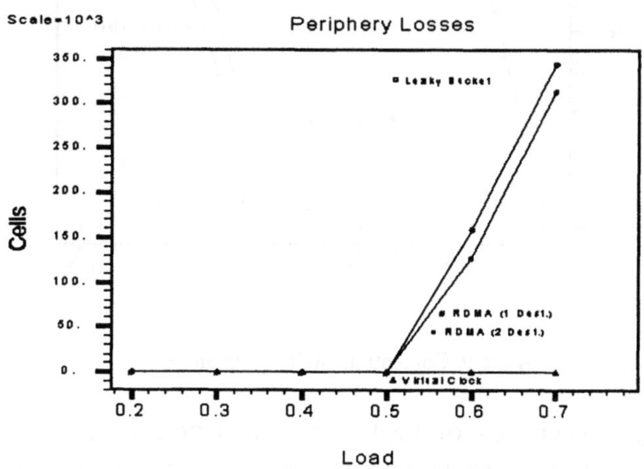

Figure 8. Loss at Backbone Periphery.

Virtual Clock is the only mechanism that, practically, does not suffer losses, even in the overload region. The possibility that a higher load (above traffic contract) enter the network could be dangerous to the other flows sharing the network when they contend at internal nodes. Therefore, Leaky Bucket and RDMA present a satisfactory result. However, Virtual Clock does not limit the extra traffic. This occurs because during periods where there is not flow check, all link capacity is available, and for virtual clock the system is stable.

The *Figure 9* depicts the results of average occupancy at backbone internal switch, C_3. This graph indicates the mechanisms effectiveness to avoid the network internal nodes be overloaded and, verifies if the mechanisms can offer a upper bound to that occupancy.

The lower average occupancy was obtained by RDMA when utilizing two routing trees, that is, two different destinations. For this configuration the active bands of RDMA at periphery are opened in intervals that do not overlap in time. Because of the opening of band at internal node is synchronized with the opening of the band at peripheral node, the average occupancy is nearly zero, even when the offered load is greater than that used to configure RDMA band size. The reason for this is that the excessive traffic is not allowed to access the backbone internal links, it is bounded at periphery. See *Figure 8*.

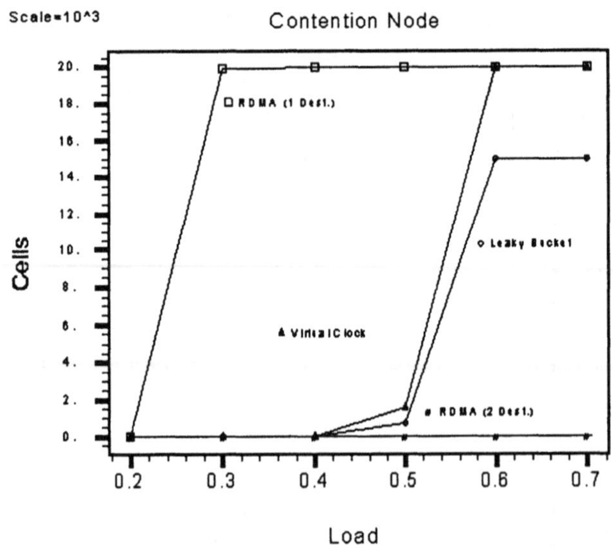

Figure 9. Contention Node Occupancy.

The average occupancy of Leaky Bucket is perceived only after the offered load becomes the same load used to its dimensioning. However, it does not grow indefinitely, being shaped as the load exceeds the contract.

Virtual Clock had a similar behavior to that of Leaky Bucket, but after a load of 50%, its mean number of discarded cells grows indefinitely as depicted in *Figure 10*.

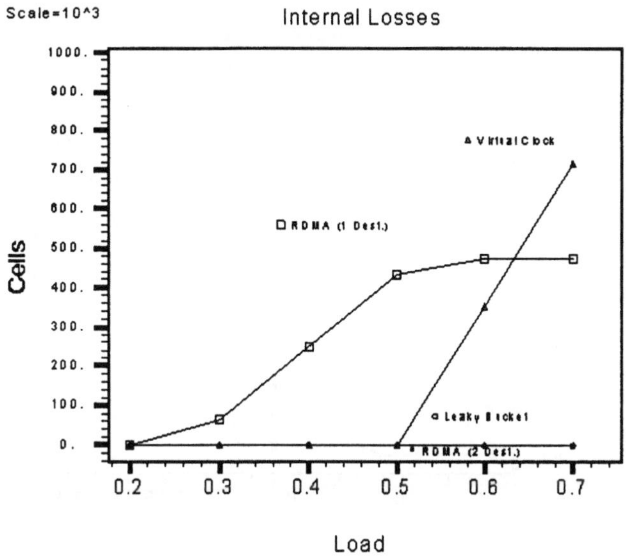

Figure 10. Contention Node Losses.

With only one routing tree (one destination for the two flows), the average occupancy of RDMA was very high, yet in the low load region. From 30%, the upper bound of internal node memory was reached and, consequently, losses occurred (See *Figure 10*). However, see that the internal RDMA node continues with only 50% of the total available cycle for this band, that is, half the link bandwidth, even now with only one routing tree. This implies that bands in peripheral nodes become active at the same time. This is an improbable situation in a RDMA routing design, that is, the sum of bands at peripheral nodes be greater than the bands at internal nodes. However, this situation is useful to show that, even using only one routing tree with overload, RDMA can limit the number of cells accessing the network as with Leaky Bucket. This can be verified in *Figure 10* where the losses for RDMA are smoothed from that load used to configure the mechanisms, that is, 50%.

3.2.2 Delay at Backbone Periphery

The experiment depicted in *Figure 11* represents the delay suffered by a MMPP flow in function of the load variation with a 95% confidence level. The MMPP is a reasonable model for this experiment because the sources have overload periods, even within the traffic contract. RDMA and the other mechanisms were dimensioned for a 50% load. For Virtual Clock we used two values for the maximal penalty time: 0.1ms (VC(0.0001s)) and 5ms (VC(0.005s)) and in both cases the time between checks (AI) was set to 0.1ms.

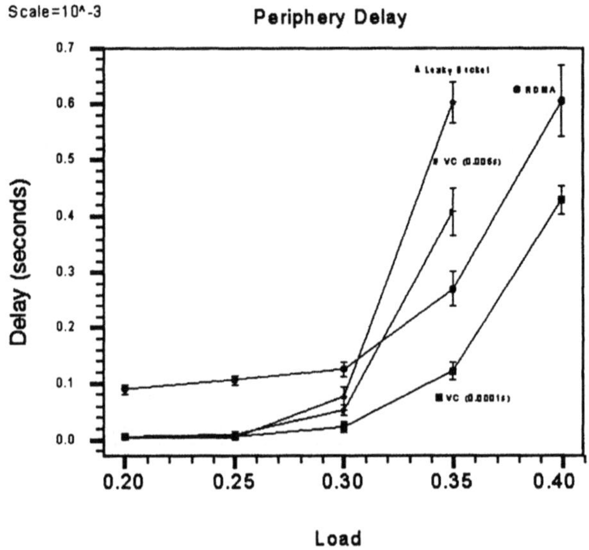

Figure 11. Delay for a MMPP Flow.

It is interesting to observe that the RDMA delay is bigger than the others for low loads, but, as the offered load grows, the delay for Leaky Bucket and Virtual Clock surpass RDMA (except for Virtual Clock with 5ms penalty time). This behavior occurs with Leaky Bucket due to the absence of credits, and for Virtual Clock due to the growing of wait queue during penalty time as load increases.

4. CONCLUSIONS AND FUTURE WORK

In this article, we have analysed a new access control mechanism for high speed networks, called RDMA (Route Division Multiple Access). We have studied the shaping effect on applications QoS requirements described in terms of delay and delay variation (jitter). We also studied the effectiveness of this new mechanism to avoid network congestion inside the network. RDMA showed itself to be satisfactory as access control mechanism for applications with strict QoS requirements, represented by delay and jitter. The traffic shaping effect, common to these mechanisms, does not cause excessive delays and jitter. RDMA seems to provide adequate performance even when it allocates resources according to the mean transmission rate.

This may present some advantages over other mechanisms like Leaky Bucket, which need to know the application burst size to effectively control the traffic.

Network internal occupancy can be controlled by RDMA. This aspect depends on the route configuration too. For flows that do not share the same routing tree internal buffer occupancy of the network can be strongly reduced. For routing trees with shared bands the utilization can be higher, but controlled. The basic difference between contention using RDMA and the other mechanisms (Leaky Bucket and Virtual Clock) is that for RDMA contention will occur only when bands share the same routing tree (destination) whereas for the others, contention occur whenever the same output link is shared. Another important aspect of RDMA is that, memories at network switching nodes may be less required than at periphery.

In sumary, RDMA can be characterized as an effective access control mechanism for real time applications. It also permits a fine tunning of congestion degree inside the network. Future reseach work includes the analysis of RDMA with more realistic traffic models, like fractal models. Another important topic is to compare the effectiveness of ATM mechanisms for admission control of new connections (CAC - Connection Admission Control) with the RDMA allocation of time bands.

REFERENCES

[1] A. Baiocchi, N. Melazzi, A. Roveri and R. Winkler. Modeling issues on an ATM Multiplexer within a Bursty Traffic Environment. INFOCOM'91, Pages 83-91, 1991.

[2] P. T. Brady. A Model for Generating ON-OFF Patterns in Two-Way Conversations. Bell System Technology Journal, 48:2445-2472, 1969.

[3] D. Florissi. Isochronets: A High Speed Switching Network Architecture. PhD. Thesis. Computer Science Department, Columbia University, 1995.

[4] B. Maglaris, D. Anastassiou, P. Sen, G. Karlsson and J. D. Robbins. Performance Models of Statistical Multiplexing in Packet Video Communications. IEEE Trans. Comm.,36(7): 834-844. July 1988.

[5] D. L. Mills. Internet Time Synchronization: The Network Time Protocol. IEEE Transactions on Communications, 39(10):1486-1493, August 1991.

[6] R. O. Onvural. Asynchrounous Transfer Mode Networks. Artech House, Second Edition, 1995.

[7] B. V. Patel and C. C. Bisdikian. End-Station Performance under Leaky Bucket Traffic Shapping. IEEE Network. 10(5):40-47, Sept/Oct.1996.

[8] S. Radhakrishnan, S. V. Raghavan and A. K. Agrawala. A Flexible Traffic Shaper for High Speed Networks: Design and Comparative Study with Leaky Bucket. Comp. Net. and ISDN Syst. 28(4): 453-469, Feb. 1996.

[9] E. P. Rathgeb, Modeling and Performance Comparison of Policing Mechanisms for ATM Networks. IEEE JSAC 9(3):325-334. April 1991.

[10]J. Rexford, F. Bonomi, A. Greenberg and A. Wong. Scalable Architectures for Integrated Traffic Shaping and Link Scheduling in High-Speed ATM Switches. IEEE JSAC, 15(5): 938-950. June 1997.
[11]M. Sidi, W.-Z. Liu , I. Cidon and I. Gopal. Congestion Control Through Input Rate Regulation. IEEE Trans. Comm. 41(3): 471-477, Mar. 1993.
[12]J. S. Turner. New Directions in Communications (or Wich Way in the Information Age?). IEEE Comm. Mag. 24(10): 8-15. Oct 1986.
[13]L. Zhang. Virtual Clock: A New Traffic Control Algorithm for Packet Switching Networks In Proc. of ACM SIGCOMM'90. Pag. 19-29, Sep. 1990.

PART FOUR

Network Programmability

Chapter 11

Inter-domain Integrity Management for Programmable Network Interfaces

Ognjen Prnjat, Lionel Sacks
Departement of Electronic and Electrical Engineering,
University College London, Torrington Place, London WC1E 7JE, U.K.
Phone: +44 20 7679 3198 Fax: +44 20 7387 4350
Email: {oprnjat | lsacks}@ee.ucl.ac.uk

Key words: Programmable networks, Parlay API, multimedia services, integrity methodology, integrity policies.

Abstract: Rapid development of the programmable network concepts anticipates a new telecommunications environment where the third party service providers will develop sophisticated user-customised multimedia applications and deploy them through the direct access to the network operators' control plane and the network itself. This is a radical change of direction, since currently operators are developing and testing the new services themselves, thus being confident of the correct and proper functioning and performance of the services, support systems and the underlying networks. This future scenario raises the issue of integrity between the network operator and the third party (service provider) domain. Here we discuss the current developments in the area of programmable networks, focusing on the open APIs such as Parlay, and present a top-down methodology for the integrity analysis and the development of integrity policies. Integrity policies give rise to the definition of integrity recommendations to be followed by network operators and service providers, or the definition of testing regimes between the two, and are ultimately enforced by the integrity-preserving mechanisms to be implemented by both parties. Moreover, the reference architecture depicting the approach for deploying the integrity mechanisms and support functionality is given.

1. INTRODUCTION

Convergence between telecoms and computing is most apparent in the field of programmable networks. Developments such as open network APIs, active networks and mobile agents pave the way for the future environment where the IT world, rich in applications, will cohabit with the telecoms world, rich in networks. By exploiting these technologies, third party application developers and end-users will have the access to the network control and infrastructure traditionally operated and run by the dominant Telcos. This scenario of open network provision is strongly encouraged by the deregulation forces in Europe and the US; and the general market trends towards rapid and efficient development and deployment of user-customised services. The services are becoming diverse, ranging from narrow-band voice telephony through broadband streams to highly dynamic multimedia application support. Demands for services are diversifying, and the requirements of availability and functionality increase as the business and industrial processes become ever more dependent on communications.

Environment is thus forming, in which the 3rd party, value-added service providers, or even end-users, will develop sophisticated applications, exploiting the infrastructure provided by the Telcos, by accessing it at the considerably lower level of control than it was previously possible. Network operators will have little knowledge of the logic of the applications deployed over their control plane, the active packets travelling through their networks, or the pieces of mobile code deployed on their equipment. However, they will require the assurance from the application developers that their applications will not harm the operation of the operators' control systems or the network as a whole. Network operators will require a guarantee that the integrity of their support systems and the network is preserved. **Integrity** is defined as [Ward95] "ability of the network to retain its specified attributes in terms of performance and functionality". The best-documented integrity breach is the January 1990 brownout [McDo94] of the AT&T American network. A control mutation originating in the switching system propagated through the signalling network causing degradation of operation and ending in a shut-down. The whole eastern seaboard of the US lost telephone connections for nine hours, with the financial loss of 1 billion dollars. A number of similar integrity breaches followed in Pacific Bell and Bell Atlantic networks in next two years [Hoan93]. Considering these incidents, the established operators rightfully fear to open their networks to less experienced players whose actions might endanger operation of the network.

Integrity issues are crucial in the future telecoms scenarios, where a number of players in the market will have to interwork, on both bearer service and management levels, while having to be confident that this interworking will not jeopardise the correct and proper functioning of the networks and systems. This paper discusses the integrity issues likely to be encountered in the future open environment. Programmable network approaches are discussed, with the focus on the open network APIs and the example Parlay API. The scenario of the inter-domain interaction between the network operator and the service provider is then discussed. Finally, we focus on the methodology for management of integrity in the inter-domain, open API-centred environments. Some integrity-preserving policies are discussed, and the reference integrity management architecture is presented.

2. INTEGRITY ISSUES

Integrity is a broad term, encompassing a variety of issues concerning system structure, functionality and behaviour. Integrity can be decomposed in a set of attributes [Prnj99a], some of which overlap with certain concepts established in the general computing fields of dependability and safety. Some basic system integrity concepts are outlined below. Note that the system is considered as a set of distributed elements providing a certain service, thus "system" and "service" can be used interchangeably.

Robustness is the ability of the system to handle unexpected events. A system that is robust can cope with all eventualities and continue operations. **Reliability** is [Reib91]: "the probability of a system performing its purpose adequately for the period of time intended under the operation conditions encountered". **Security** refers to the ability of the system to detect and avoid intentional external attack. **Resilience** is the ability of the system to recover from faults. **Availability** is "the percentage of time during which the system is operational and conforms to its specification" [Vill91]. **Performance** is the throughput of the system, often traded off against functionality. **Scalability** is the impact on *performance* as more entities (processes, devices *etc.*) are added to the system. **Data Coherence**: the information copied or distributed through the system needs to remain consistent through time and change of circumstances. **Liveness** means that something will, eventually, happen: a system might not remain live because it is in a state of either *deadlock*, when it is expecting an event which cannot occur; or *livelock*, when it oscillates between a closed set of states that it cannot leave. **Feature Interactions** refer to when two or more systems, each with well-defined and understood behaviour, result in unforeseen behaviour while operated together [Came93]. **Complexity** may be an assessment of how long (how many iterations or cycles) an operation takes – known as ***time* complexity**. ***Computational* complexity** relates to how well a given procedure can be analytically described or determined. ***Data* complexity** refers to the complexity of data structures and their interdependencies. A high level of complexity, unless it is there to increase robustness, poses more threats to system operation and thus to integrity. A high level of **coupling** between system components indicates a high level of interdependence: a failure, or an integrity breach, can ripple through the system via the coupling paths. A distinct sub-attribute of integrity is **risk**: "a measure that combines both the likelihood that the system hazard will cause an accident and the severity of that accident" [Std1228-94]. Here we view risk essentially as ~ ***1/integrity***, *i.e.* the higher the risk, the lower integrity.

The topics described above are all interrelated. For example, a system with poor scalability will loose performance as the number of entities it involves grows. This would reduce availability. Loss of performance and availability can impact on timing of dependent systems, by impacting on data coherence or the liveness constraints. Thus, a framework for analysing the integrity issues and building the desired level of integrity into newly developed systems and applications is of crucial importance. This is particularly important in the future scenarios of programmable networks where the network operators will have no notion of what are the integrity impacts of opening their control and networks to the third parties.

3. PROGRAMMABLE NETWORKS

There are numerous approaches to programmable networks. The initial *active network* concepts are based on data packets which travel through the network carrying programs in their header which can run on network devices such as switches or routers [Tenn96]. These concepts progressed through active bridging [Alex97], where packets carry only the flag indicating the desirability of running a program. The latest initiative in this area is the application-level active networking [Mars99]. *Mobile agents* are another big arena of research and development, telecoms [Karm97] applications including IN-based service provisioning, network management by delegation [Gold98] and personal communication services. Finally, the *network programming interfaces,* on which this paper concentrates, provide an open access for service developers to the service components and network control in the operator domain [Laza97]. Using these interfaces, service providers can develop their own applications using the underlying control and network infrastructure provided by the operator.

Main initiatives in the area of programming interfaces are those led by the IEEE Project P1520, which aims at standardising the programming interfaces for networks [Bisw98], and the industry-led Parlay group, which is specifying and developing a new open network application programming interface [Parl00]. Parlay specification provides an object-oriented open interface, the Parlay API, to the network generic services in the operator domain. P1520 specifications are somewhat broader, encompassing the CCM interface to the physical element level, the L-interface to the virtual network device level, the U interface to the network generic service level, and the V interface to the value-addedd services level, while focusing mostly on the L interface and to some extent the U interface.

In this paper, we focus on the integrity of interfaces between autonomous domains, that is, the network operator domain and the 3rd party, service provider domain. This interface is referred to as the U interface in the P1520 reference model, and the Parlay API in the Parlay reference model. Common feature of these interfaces is that they essentially specify the interconnection point between the two autonomous domains: network operator and service provider domains. These interfaces can be compared to the ConS and Xuser interfaces in TINA [TINA-BM] and TMN [M.3010] architectures.

3.1 Parlay inter-domain scenario

Parlay API is a typical example of a network programming interface. Phase 1 Parlay specification defines an interface for call control (PSTN) and messaging (email, voice), while phase 2, released in January 2000, provides for wireless and IP services. Parlay API is object oriented, extensible (in terms of functionality, management and billing), and network technology independent, providing the user with application portability.

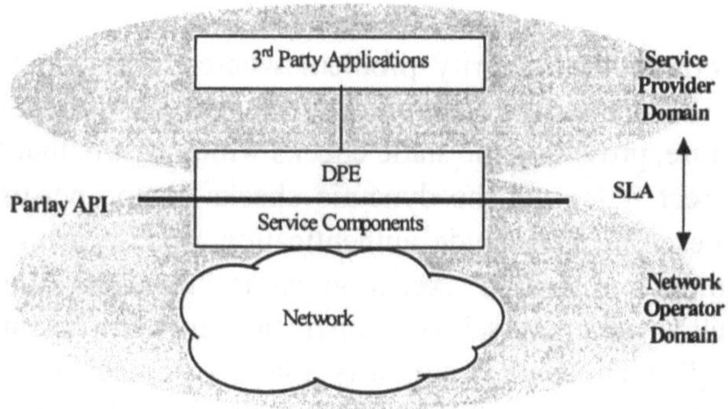

Figure 1. Parlay inter-domain scenario

The context of the use of the Parlay API in the inter-domain interactions is shown on Figure 1. The ellipses represent separate administrative domains. Network operator operates the network resources and the specific service components such as configuration, routing, billing, storage, *etc.* in its own domain. Operators' service components, and through them the network itself, are accessible via the Parlay API. 3rd party service applications access this API, from the service provider domain, through a Distributed Processing Environment (DPE), which can be CORBA, JAVA RMI, DCOM or other.

The problem in this context is the preservation of the integral operation of the operators' domain, which can be at stake due to the degree and logic of the interaction of the 3rd party applications with the service components and sensitive network resources. A number of applications will directly interfere with the network operation, and they may do so with variable frequency and load. Moreover, the network operator has no knowledge of the applications logic and thus can not effectively manage the integrity of its own domain. First step towards the solution to this problem is the establishment of a Service Level Agreement (SLA) between the two players. SLA here has a somewhat broader scope than that of a service conflict resolution tool between the two parties, for what it is mostly used nowadays. SLA is envisaged to embrace the service usage definitions and integrity requirements as well. Considering the issues of service usage, SLA would typically define what operations is the service provider allowed to perform, and the expected resource usage. However, the SLA would also need to define the agreement between the operator and the service provider considering the pure integrity issues, discussed in section 2. In order to successfully define the integrity issues to be included in the SLA, and develop the ways of ensuring the integrity of the interaction, both players will have to follow an integrity management process.

3.2 Existing integrity management approaches

Some solutions to the integrity problem where offered in the area of active networks. Secure Active Network Environment (SANE) [Scot98] is a layered architecture, providing the static checks which ensure that the system starts in the correct state, and the dynamic checks on the per-user or per-packet basis. These checks include authentication, and securing the active node by providing a restricted execution environment for the 3rd party programs, as well as a naming scheme to partition the node resources. Proof Carrying Code (PCC) [Necu97] similarly focuses on the integrity of the active node, by attaching a proof of correctness to each active packet, which is checked at the node before program execution. Finally, a number of active network programming languages, such as Safetynet [Wake] and PLAN [Hick99] aim to ensure integrity at the level of the programming language by tightly controlling the processing time, memory allocation and threading, removing the requirements for run-time checks, and similar.

These solutions seem effective for the protection of the active nodes themselves, but are not fully applicable in the Parlay scenario where the programmability is taking place on the interface to the control plane, rather then on the element level. In what follows we present an integrity management methodology and a reference management architecture targeted at the preservation of integrity in the scenario of the open network interfaces.

4. INTEGRITY MANAGEMENT

4.1 Methodology

The integrity management methodology presented here is based on the methodology introduced in [Prnj99a]. Three phases in this methodology are prediction, testing and maintenance. **Prediction** is a pre-emptive activity, assessing the features and the overall system integrity status prior to launching. It aims at locating and removing integrity risk areas - hotspots - during system development, and building in the integrity preserving mechanisms. **Testing** is conducted during system development, and as the final phase before the system launch. Testing here refers to the final stand-alone tests, and tests performed during integration, which are of particular relevance in multi-domain environments. **Maintenance** is conducted after system deployment and aims at detecting any malfunction or degradation of operation that might pose a risk to system integrity. It involves the measurement of integrity-relevant features of the operational system, diagnostics of cause of degradation, and application of the relevant response.

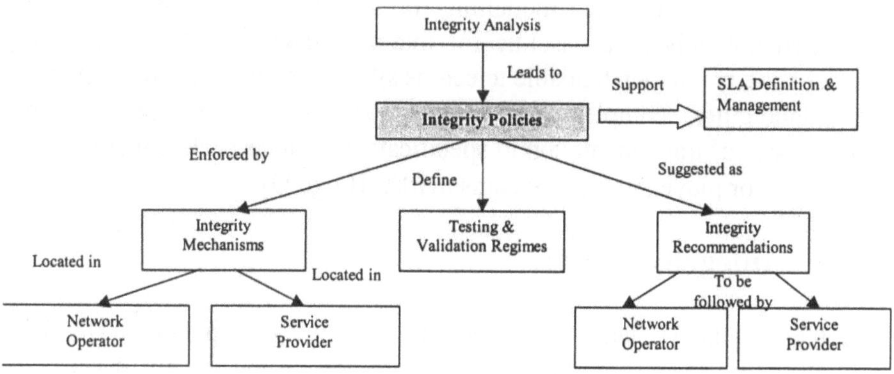

Figure 2. Integrity methodology

Prediction phase is based on the iterative conduction of integrity analysis, and development and implementation of integrity policies throughout the system/service development lifecycle. Firstly, analysis of the 3rd party application is conducted, in terms of its own structure and in terms of the interaction with the network operator domain. Integrity analysis takes the issues discussed in section 2 into account, at different stages of development lifecycle. Based on analysis, integrity policies, aimed at preservation of integrity of the interaction between the two domains, are defined. Integrity policies can give rise to definition of the integrity recommendations to be followed by network operators and service providers, or definition of testing regimes between the two (thus leading to the testing phase of the methodology), and are ultimately enforced by the integrity-preserving mechanisms to be implemented by both parties. The mechanisms enforce the policies, and thus belong to the maintenance phase of the methodology. Policies support the SLA definition and management (*Figure 2*).

Integrity analysis considers a number of integrity issues discussed in section 2. Depending on the stage in system/service development lifecycle, these issues can be analysed either through the use of semi-formal modelling techniques (such as Unified Modelling Language, UML [UML]), or more formal behavioural simulation techniques such as Specification and Description Language, SDL [Z.100]. Detection of the high-risk components early in the development lifecycle can be accomplished using the system metrics [Prnj99b], which pin-point components with high complexity and coupling levels. In its own part, behavioural modelling is of critical importance since the Parlay API specification does not include any behavioural properties, and thus is not robust in terms of operation invocation sequencing and timing that can bring the service components and the network in the undesirable state, causing deadlock or livelock.

Semi-formal and formal modelling for system development is strongly advocated not only as a baseline for integrity analysis. Formal modeling ensures coherent, understandable telecoms system development by providing for unambiguity, completeness and consistency of specifications, and conformance of implementation to specifications; and as such is increasingly used by major players in the telecoms market [Loge99].

4.2 Integrity policies

Integrity analysis leads to the definition of integrity policies. These in turn lead to design recommendations to be followed by parties in autonomous domains. They also specify testing and validation regimes, and are ultimately enforced by the integrity preserving mechanisms in the appropriate domains. Some integrity policies can be a responsibility of the DPE used to support the interaction. DPE can sometimes provide for naming, timing, transaction (for preservation of data coherence) and security support. Integrity policies can be diverse, depending on the output of the integrity analysis. Some integrity policies are discussed in what follows.

Typical policy that takes a form of a design recommendation is the use of **behavioural techniques** in specifying not just the logic, *i.e.* the behaviour of the application, but also the behaviour of the API offered, in the operator domain. Current Parlay API specification is flat: it has the form of UML class diagrams and shows neither the behaviour of the objects implementing the interface, nor the relationships between objects. Some simple UML state diagrams describe the behaviour of few complex objects, and sequence diagrams are specified to guide the developers. This is not enough for specifying and enforcing the correct sequencing of method invocations on the API, and cannot guarantee the preservation of liveness. State diagrams defined for the objects implementing the API interfaces have to be carefully designed and behavioural modelling has to be fully exploited [Kolt00], including reachability analysis, livelock/deadlock detection and robustness.

Data coherence policies support atomicity, consistency, isolation and durability of transactions on data. Example of such integrity policy mechanism is the Transaction Service for network management applications [Ranc98]. This could be the shared responsibility of the DPE.

Policy requiring liveness would employ techniques to detect and resolve any livelock/deadlock states reached by the objects implementing open interfaces. This would typically be the responsibility of the operator. Similarly, integrity mechanisms detecting and resolving **feature interactions** could be deployed in the operator domain.

Security mechanisms can be used to implement **security policies** in both domains, dealing with authentication, access control, data integrity, confidentiality and non-repudiation. Current Parlay API Core Specifications provide only the authentication mechanisms through the framework interface. An example [Psou99] security policy would be a reference monitor based on access lists, restricting the access to the service, information, and system resources in the operator domain. This could be further enhanced with limitation techniques for the use of resources in terms of memory, number of subprocesses, CPU time or other. Note that the automatic establishment of the SLA must also be secure: the digital signature (confidentiality and non-repudiation) would be required.

Integrity mechanisms monitoring for **performance** in terms of throughput and delay can patrol for the compliance with the possible performance-related aspect of the SLA agreement.

Some integrity policies discussed above could be enforced using the authorisation/obligation policy approach as defined in [Slom99]: authorisation policies define what operations can be performed on target objects, and obligation policies define what a manager (in our case the 3rd party application) must do as a response to an event trigger. However, some integrity policies have more dynamic properties, in terms of timing, sequencing, performance *etc.*, and are not fully describable using the pre-condition, parameter and timing constraints defined.

4.3 Reference integrity architecture and mechanisms

The reference architecture of Figure 3 depicts the proposed integrity mechanisms which enforce the integrity policies in both service provider and network operator domains; and the additional support integrity functionality. Figure 3 illustrates the same scenario as Figure 1: the two domains are now shown communicating through the API over the DPE. The DPE provides the basic services, like transaction, naming and security (details are dependent on the DPE used). Integrity-supporting elements in the service provider domain include an Integrity Framework (FW) as well as operations and management interfaces (MS) to the operator domain. Operator domain contains the Service Control Point (SCP) for the interface with the control plane, Network Service Emulator (NSE) and Validator (NSV) systems, and the integrity gateway (IG).

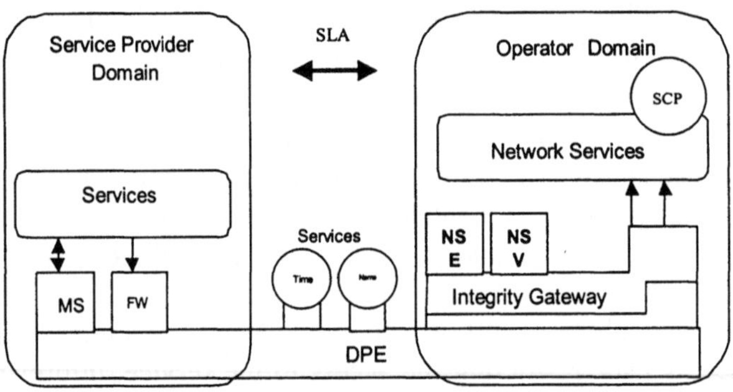

Figure 3. Reference architecture

The **integrity gateway** (IG) in the operator domain is the key component implementing the integrity mechanisms and thus enforcing integrity policies. Additional policy enforcement is provided by components in the service provider domain - the integrity framework (FW). The objective of the access integrity gateway is to effectively narrow the state and parameter space of requests being made over the open interface to the operator's service platform. The gateway has the characteristics of a service access server by giving access to network services and resources and monitoring for integrity violations. It should provide not only integrity enforcement, but also trouble-shooting, violation alarms, event reports and diagnostics. Examples of integrity mechanisms implemented in the gateway would be, as discussed in section 4.2, components that monitor system dynamics for livelock or deadlock, data coherence, feature interactions, performance degradations, parameter range violations, resource miss-allocations and such like. Each of these may be dependent on the contractual issues such as: which DPE and DPE services are to be used; which services are being offered, at what successful transaction rate and at which level of robustness. These aspects of the provided service are captured in the SLA.

Some integrity policies give rise to the definition of validation and testing regimes (section 4.1). An overall test environment is a central requirement for the support of many (principally pre-deployment) integrity issues. Focus of this environment is a network **service emulator** (NSE, Figure 3) capable of taking the place of the actual network control system. This emulator would aid development of service environments and the integrity gateway, without requiring access to a control platform. Further, the emulator would enable services to be tested and validated against integrity requirements prior to deployment. This testing is not envisaged to be exhaustive and delay service roll-out, but tailored to the particular integrity requirements of a particular interaction scenario. In this role the emulator would be developed as a network **service validator** (NSV, Figure 3) and would constitute a clear specific component of the integrity maintenance system. In complement to the validator and the service emulation, a set of service provider-side test suites should be developed to perform pre-production integrity tests on the operator's service provisioning platform. In this way, development of the service emulator not only aids in the development of integrity mechanisms, but establishes a mechanism for testing both the general open network control environments and specific service developments.

4.4 Inter-domain SLA management

Different policies for integrity management were discussed in this paper. Depending on the actual scenario of the inter-domain interconnection on the level of the programmable network API, different policies would need to be applied. Policy requirements are specified within the integrity-enhanced SLA established between the two parties in autonomous domains. These requirements would possibly be parametrised, for example as threshold criteria in the form of: "if a certain integrity-related parameter exceeds x, than it is considered as a threat and thus is not acceptable" [Ward95].

Target policies need to be expressed through a formal language. This would ensure the exact and common understanding of the policies by the parties involved in the interconnection, and the automated way of interpreting and enforcing these policies in the integrity gateway. Addition of the axioms and inference rules to the formal policy language would yield a formal logic capable of detecting policy conflicts (for some approaches consult [Lupu97]) which may arise due to the inconsistent policy specifications when multiple integrity policies are defined. XML [XML] is being considered as a possible policy language.

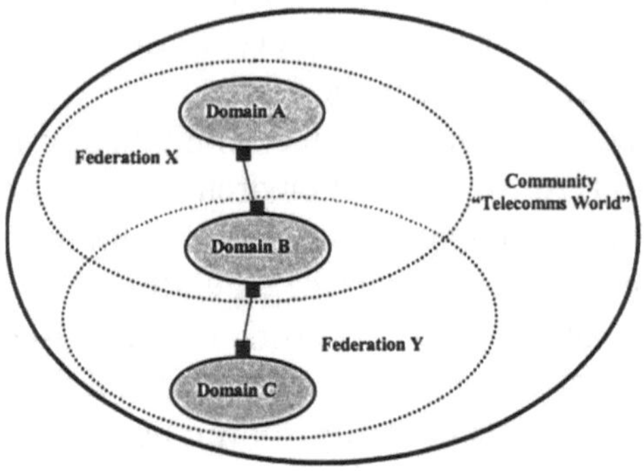

Figure 4. SLAs in federated environments

In the federated environments consisting of multiple autonomous players, the target integrity policies can be established during development and for testing, and the SLAs could be exercised prior to the deployment of 3rd party applications. This approach is more lax in the sense that the integrity policies and SLAs can be established solely by parties willing to indulge in the interconnection. Alternatively, global, regulatory conditions based on quantitative integrity/risk notions can be established by the standardisation bodies and imposed by the regulatory bodies. These two approaches are shown on *Figure 4*. Domains (autonomous players) are marked as dark ellipses. Federations are marked as grey ellipses, representing groupings of autonomous players with a common objective. Community representing the whole telecoms community is marked by the hollow ellipse. First approach discussed advocates the use of different SLAs (marked by black lines) between domains that form a federation. Federation is based on a certain SLA - a set of integrity rules defining design recommendations, testing regimes, and run-time integrity mechanisms to be deployed. Alternatively, global integrity policies can be established in the telecoms community, governing inter-domain SLA definitions between any two parties.

5. CONCLUDING REMARKS

This paper identified some general integrity issues that must be considered in the context of future telecoms scenarios where the networks and control systems traditionally managed by the dominant Telcos will be more open to the 3rd party service deployment. An integrity management methodology was presented paving the way for the network operators and third party, value-added application developers and service providers to understand and manage the integrity issues. Focus of the methodology was on the programming network interfaces, such as the Parlay API. The methodology gives rise to definition of: integrity recommendations to be followed by the network operators and application developers; testing and validation regimes to be deployed between the autonomous domains; and implementation of integrity-preserving mechanisms enforcing the integrity policies in both domains. A conceptual reference architecture was presented, encompassing two main integrity enforcing components: integrity gateway in the operator's domain, and the integrity framework in the service provider domain; as well as the service emulator and validator components in the operator domain supporting pre-launch service emulation and validation. The proposed integrity methodology focused strongly on the programming network interfaces, *e.g.* the Parlay API, and as such is inevitably operator-centred. Most of the integrity issues in the open API scenarios are solely manageable by the operator. Although existing techniques such as Proof Carrying Code tackle the integrity problem to some extent and put some of the integrity management responsibility on the application developer, they do not resolve issues such as feature interactions, livelock and deadlock detection *etc*. Authors thus believe that the approach proposed in this paper, although relatively heavy-handed, represents a viable solution in the under-researched field of the programmable interfaces integrity management.

ACKNOWLEDGEMENTS

Authors would like to thank: Frans Panken, Lucent Technologies; Jeremy Wilson and Martin Yates, BT Labs; Rob Brennan, TelTec DCU; and Ina Schieferdecker, GMD Fokus; for their help in defining some ideas expressed in this paper. This work is funded in part by the EPSRC grant M26091.

REFERENCES

[Alex97] Alexander et. al., "Active Bridging", Computer Communications Review, 27, 4, pp. 101-111, 1997.

[Bisw98] J. Biswas et. al., "The IEEE P1520 Standards Initiative for Programmable Interfaces", IEEE Communications Magazine, October 1998., pp 64-72.

[Came93] E. J. Cameron and H. Velthuijsen, "Feature Interactions in Telecommunications Systems", IEEE Communications Magazine, August 1993, pp. 18-23.

[Gold98] G. Goldszmidt and Y. Yemini, "Delegated Agents for Network Management", IEEE Communications Magazine, March 1998, vol. 36, no. 3, pp. 66-70.
[Hick99] M. Hicks et. al., "A Secure PLAN", Proceedings of the First International Working Conference on Active Networks (IWAN '99), Published by Springer-Verlag.
[Karm97] A. Karmouch, "Mobile Software Agents for Telecommunications", IEEE Communications Magazine Guest Editorial, vol. 36, no. 7, 1997.
[Kolt00] M. Koltsidas, O. Prnjat, L. Sacks, "Development of Parlay-based Applications Using UML and SDL", MMNS' 2000, September 2000.
[Laza97] A. Lazar, "Programming Telecommunication Networks", IEEE Network, October 1997, pp. 2-12.
[Loge99] X. Logean et. al., "On Applying Formal Techniques to the Development of Hybrid Services", IEEE Communications Magazine, Vol. 37, No. 7, July 1999, pp. 132-138.
[Lupu97] E. Lupu and M. Sloman, "Conflict Analysis for Management Policies", Integrated Management IM '97 Conference Proceedings.
[M.3010] ITU Rec. M3010, "Principles for a Telecommunications Management Network".
[Mars99] I. W. Marshall, et. al., "Application-level Programmable Network Environment", BT Technology Journal, Vol. 17, No. 2, April 1999.
[McDo94] J. C. McDonald, "Public Network Integrity - Avoiding a Crisis in Trust", IEEE Journal on Selected Areas in Communications, January 1994, vol. 12, no. 1.
[Necu97] G. C. Necula, "Proof-Carrying Code", 24th ACM SIGPLAN-SIGACT Symposium Proceedings, 1997.
[Parl00] The Parlay Group, "Version 2.0 Parlay Specification", http://www.parlay.org/
[Prnj99a] O. Prnjat, L. Sacks, "Integrity Methodology for Interoperable Environments", IEEE Communications, May 1999, Vol. 37, No. 5, pp 126-139.
[Prnj99b] O. Prnjat, L. Sacks, "Telecommunications System Design Complexity and Risk Reduction Based on System Metrics", EWDC10, Proceedings published by OCG, 1999.
[Psou99] K. Psounis, "Active Networks: Applications, Security, Safety and Architectures", IEEE Communications Surveys, First Quarter 1999.
[Ranc98] D. Ranc et al., "Use of Transactions in Network Management Applications", Intelligence in Services and Networks, IS&N'98 Conference Proceedings.
[Reib91] A. L. Reibman, M. Veeraraghavan, "Reliability Modelling: an Overview for System Designers", IEEE Computer, April 1991.
[Scot98] D. Scott et.al., "A Secure Active Network Environment Architecture", IEEE Network, May/June 1998, vol. 12, no. 3.
[Slom99] M. Sloman, E. Lupu, "Policy Specification for Programmable Networks", Proceedings of IWAN '99 proceedings, Published by Springer-Verlag.
[Tenn96] D. Tennenhouse, D. Wetherall, "Towards an Active Network Architecture", Computer Communications Review, 26, 2, 1996.
[TINA-BM] H. Mulder (Ed.), "Business Model and Reference Points", Version 4.0, www.tinac.com/specifications/specifications.htm
[UML] Rational Software Corporation, Unified Modelling Language, www.rational.com/
[Vill91] A. Villemeur, "Reliability, Availability, Maintainability and Safety Assessment, Vol. 1" J. Wiley, 1991.
[Wake] I. Wakeman et.al., The Safetynet Project, www.cogs.susx.ac.uk/projects/safetynet/
[Ward95] K. Ward, "Impact of Network Interconnection on Network Integrity", British Telecommunications Engineering, January 1995, vol. 13, pp. 296-303.
[XML] Extensible Markup Language (XML) W3C Recommendation 10.2.1998.
[Z.100] ITU Rec. Z.100, "Specification and Description Language (SDL)", 1993.

Chapter 12

A TINA-based Distributed Environment for Mobile Multimedia Applications

Alexandre S. Pinto, Luis F. Faina and Eleri Cardozo
DCA – FEEC – UNICAMP, PO Box 6101
Campinas, SP, Brazil, CEP 13083-970

Key words: TINA, DPE, CORBA, Mobile Systems, Multimedia Systems

Abstract: This paper describes the design and implementation of a component of the Telecommunications Information Networking Architecture (TINA): the Distributed Processing Environment (DPE). TINA provides concepts, principles, models and standards for the next generation of telecommunication services. These new services take advantage of modern technologies such as high speed networks, multimedia processing, distributed objects, and component-oriented software development. The goal is to devise services that can be introduced, modified and withdrawn as soon as market demands are identified. TINA DPE provides the infrastructure necessary to distribute the service components among service users, providers and retailers. Our DPE implementation offers two basic facilities to the TINA applications: life-cycle and stream facilities. Life-cycle facilities allow distributed objects be deployed and managed transparently, while stream facilities allow application objects to exchange media flows such as audio and video flows.

1. INTRODUCTION

From the last two decades the telecommunication market is experiencing a dramatic growth pushed by demands from large enterprise business to hobbyists at home [1]. However, the current telecommunication infrastructure was designed to carry voice and data, not to provide sophisticated services. The improvement of existing services is currently limited by the lack of computing power available at the switching facilities and central office. As a result, the introduction of a new service or modification of an existing one is a slow, costly and labor intensive process.

Back in 1992, with this scenario in mind, about forty companies including telecom operators and equipment suppliers plus computing manufacturers and software suppliers, founded the Telecommunications Information Networking Architecture Consortium – or TINA-C [2]. The chief objective of this consortium is to devise an open architecture (the TINA architecture) for supporting the development and management of telecommunication services, no matter how complex a service would be. In the TINA architecture services are pieces of software with components distributed among users and providers. Being software-based with high degree of platform independence, object-oriented (or more recently, component-oriented), and based on well accepted industry standards, services can be rapidly introduced, adapted and withdrawn according to the market demands.

This work describes the design and implementation of a component of the TINA architecture: the Distributed Processing Environment (DPE). The DPE is an infrastructure responsible for the distribution of the software implementing a telecommunication service. Our DPE implementation is based on CORBA[1] [3] and RM-ODP[2] standards [4]. The highlights of this implementation are the supporting for multimedia flows and object migration.

This paper is organized as follows. Section 2 presents a short introduction to the TINA architecture. Section 3 provides details of the design and implementation of a CORBA-compliant TINA DPE. Section 4 presents how multimedia streams are supported by the DPE as well as how this facility is integrated with the life-cycle facilities. Section 5 presents some implementation issues and an application built above our DPE. Finally, section 6 closes the paper with some concluding remarks and future research directions.

2. OVERVIEW OF TINA

TINA [5] is a software architecture targeted to telecommunication systems and designed according to two principles: distribution and object-orientation. Distribution means that TINA software is decomposed into a set of components, being these components located at different processing nodes. Object-orientation states that the unit of distribution is an object or an aggregation of objects.

[1] Common Object Request Broker Architecture from the Object Management Group (OMG).
[2] Reference Model for Open Distributed Processing from ISO.

The TINA architecture is depicted in Fig.1. Each processing node is based on some hardware architecture that runs a Native Computing and Communication Environment, the NCCE. NCCE can be viewed as a dedicated or general-purpose operating system plus a stack of network protocols. Each processing node has an implementation of the DPE running on it.

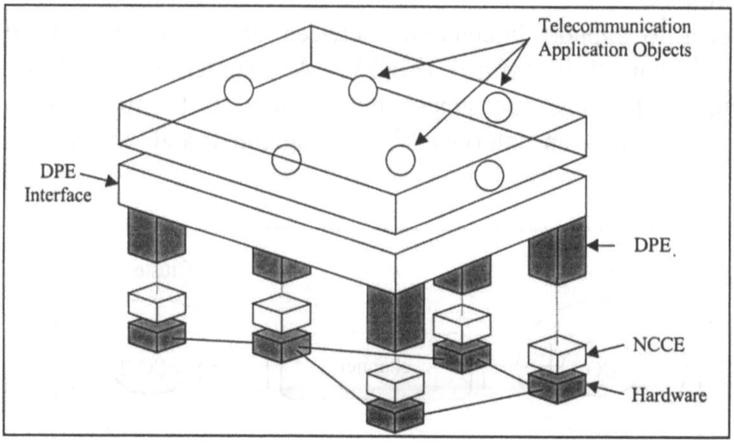

Figure 1. TINA Environment [5]

The DPE offers a homogeneous interface to the TINA applications (Fig.1). Some DPE functionalities may be missed at some NCCE. For instance, a NCCE dedicated solely to management tasks can be deprived from multimedia stream facilities.

The TINA DPE [6] has two major functions: distribution (deployment) and communication. The distribution function defines the unit of distribution as *engineering computational object* - eCO. An eCO has one or more interfaces. eCOs are instantiated from *templates*, something similar to classes in object-oriented programming languages. A template states rules for object instantiation such as the initial state and interfaces. For management purposes, objects are aggregated into *clusters*. A cluster is a unit of spatial location, activation/deactivation and migration. eCOs can be created and removed from a cluster during the cluster's lifetime. Clusters live inside *capsules*, the TINA's unit of resource allocation. An interesting (and rarely implemented) property of clusters is migration: a cluster can migrate from a capsule to another carrying all of its objects. The last component in the deployment model is the *node*, a unit of network connectivity and network management. A node provides DPE support and manages its computing resources autonomously.

In order to interact, objects must bind their interfaces. The DPE communication function defines two types of interfaces: operational and stream interfaces. The interaction that occurs at an operational interface is structured in terms of invocations of one or more operations, and possibly responses to these invocations. In a stream interaction, the information exchange occurs in the form of unidirectional media flow (e.g., audio and video flows).

In order to support interactions between objects located at different clusters, a channel is defined (Fig.2). A channel is an infrastructure providing mechanisms for supporting distribution transparency (access, location, etc.) plus an interface for management and control purposes (Fig.2).

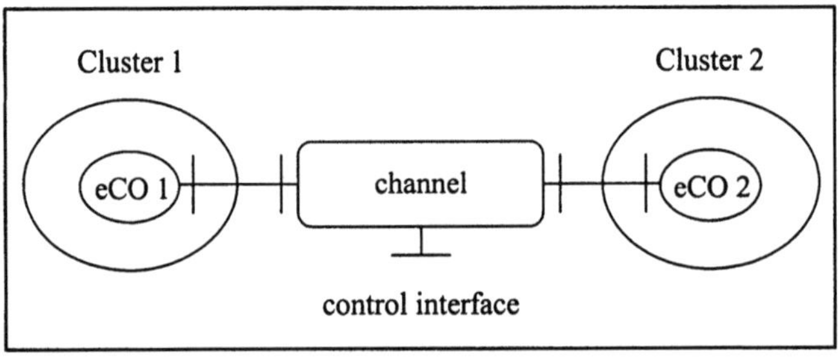

Figure 2. A channel binding two eCOs

3. DPE ARCHITECTURE

The architecture of the implemented DPE is shown in Fig.3. The DPE is built over a CORBA platform [7]. A CORBA platform allows interactions between objects in a distributed heterogeneous system. DPE instances are connected through the OMG's Internet Inter-ORB Protocol (IIOP), a standard for ORB interworking [3].

The two main facilities available at our DPE are the stream [8] and the life-cycle [9] facilities. Stream facilities allow the telecommunication services to manipulate multimedia streams (flows), while life-cycle facilities allow the service components to be distributed and managed across the network. Stream facilities are fully compliant with the OMG's *Control and Management of Audio/Video Streams* [10] and the life cycle facilities were defined according the TINA DPE specifications [5] [6].

Back to Fig.3, the life-cycle facilities provide functionalities for the management of eCOs, clusters and capsules (including cluster migration), and the stream facilities provide functionalities for the establishment and management of channels. All the DPE functionalities, available at the DPE interface, are accessed through an object request broker (ORB).

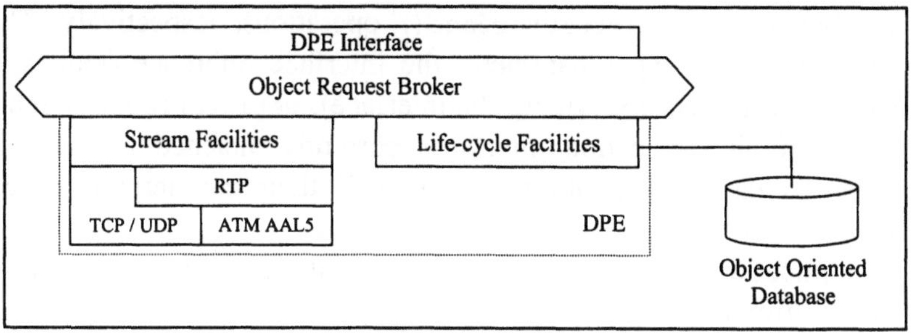

Figure 3. DPE Architecture

3.1 Support for Life-Cycle

The support for life-cycle in our TINA DPE is realized through management interfaces at object, cluster, capsule and node levels.

Each interface defined in a TINA object is realized through a CORBA object. A special CORBA object, the eCO manager, defines an interface for management purposes (the remaining interfaces are referred as application interfaces). This management interface has the following IDL description:

```
interface eCOManager {
  long addInterface(in Object interface_ref);
  long removeInterface(in Object interface_ref);
  sequence<Object> getInterfaces();
  long checkpoint(in string template);
  long recover(in string template);
  long Delete();
  long deactivate(in string template);  };
```

The operation `checkpoint()` saves the state of the CORBA object in a template; `recover()` restores the state of the CORBA object from a template; `Delete()` acts as a destructor for the CORBA object, releasing all the resources allocated to the object; and `deactivate()` is equivalent to a `checkpoint()` followed by a `Delete()`.

Typically, templates point to database entries that store the state of objects in persistent memory. The `eCOManager` interface performs management functions within a TINA object. This interface is application independent, being its implementation provided by the DPE.

Application interfaces are added to and removed from an eCO via `addInterface()` and `deleteInterface()` operations, respectively. These operations receive as input parameter the interface reference (that is, the CORBA object reference servicing the interface). `getInterfaces()` returns a sequence of all application interfaces currently attached to the TINA object. The remaining operations simply call their counterparts at each CORBA object servicing the application interfaces.

The application is responsible also to provide a factory for object instantiation from templates.

The management of clusters is provided by cluster managers implementing the following IDL interface:

```
interface ClusterManager {
  eCOManager makeObject(in string object_name);
  sequence<eCOManager> getObjects();
  long checkpoint(in string template);
  long recover(in string template);
  long Delete();
  long deactivate(in string template);  };
```

The operation `makeObject()` installs a new object in a cluster, returning a reference of its eCO manager. The input parameter is a name from which the object is identified in the cluster. This new object is created without any application interfaces. `getObjects()` returns the references of the object managers in the cluster. The remaining operations, `checkpoint()`, `recover()`, `deactivate()` and `Delete()` call their corresponding operations at each object manager in the cluster.

The capsule management interface is given below:

```
interface CapsuleManager {
  ClusterManager makeCluster(in string cluster_name);
  sequence<ClusterManager> getClusters();
  long reactivate(in string cluster_name, in string template);
  long migrate(in string cluster_name, in string capsule_name,
in string node_name);  };
```

The operation `makeCluster()` creates a new empty cluster within the capsule, returning a reference of its cluster manager. The input parameter is a name from which the cluster is identified in the capsule. `getClusters()` returns the references of the cluster managers in the capsule. `reactivate()` installs a new cluster from a template generated during a cluster checkpoint. Figure 4 illustrates a migration scenario accomplished through deactivation followed by reactivation of the migrating cluster.

Node management is performed by node managers implementing the following IDL interface:

```
interface NodeManager {
  sequence<CapsuleManager> getCapsules();   };
```

This interface defines only one life-cycle operation, `getCapsules()` that returns all the capsules created within the node.

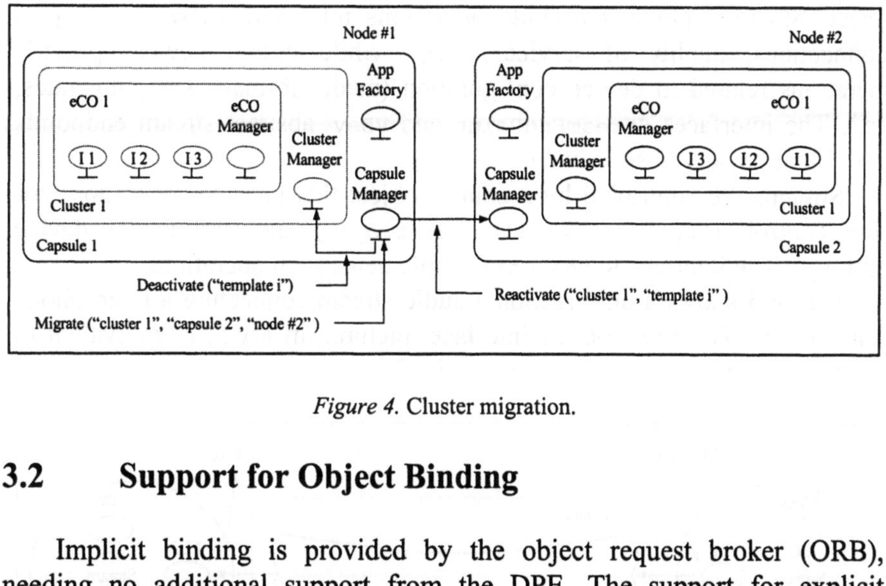

Figure 4. Cluster migration.

3.2 Support for Object Binding

Implicit binding is provided by the object request broker (ORB), needing no additional support from the DPE. The support for explicit binding through channels is an important component of the DPE. As stated before, in our implementation, channels are realized above an infrastructure based on the OMG standard *Control and Management of Audio/Video Streams* (AVStreams) [10]. This infrastructure allows the establishment and management of media flows, and its functionalities are related to the channel control interface shown in Fig.2. AVStreams, as detailed in the next section, is fully integrated with the life-cycle facilities. This means, for instance, that when an eCO is destroyed, the channel endpoints connected to it are also destroyed. Also, when an eCO migrates with its cluster, the channel endpoints are reconstructed in the cluster's new location.

4. DPE STREAM FACILITIES

Streams are facilities for supporting continuous media transfer. A stream aggregates a set of *flow connections* and terminates on a *stream endpoint*. Each flow connection is unidirectional and ends on a *flow endpoint* (FEP) within a stream endpoint. Flow endpoints can be a flow consumer or flow producer.

Multimedia devices abstract physical or logical devices that generate or consume media segments. Multimedia devices are bounded through streams in order to establish a media flow between a media producer and one or more media consumers. The interface `MMDevice` defines operations related to multimedia devices.

Inside multimedia devices two objects are responsible for controlling the media transferring: the stream endpoint object and the virtual device object. Stream endpoints provide operations related to network transport (connections, quality of service, etc.), while virtual devices provide operations related to device configuration (media formats, sampling rates, etc.). The interfaces `StreamEndPoint` and `VDev` abstract stream endpoints and virtual devices, respectively.

Streams are controlled by a control object that provides operations for flow control (`start()`, `stop()`, `destroy()`) and multimedia device bindings. The interface `StreamCtrl` implements such operations.

Figure 5 shows a unidirectional audio stream connecting a microphone to a speaker. The `StreamCtrl` interface, pictured in Fig.5, is equivalent to the channel control interface shown in Fig.2.

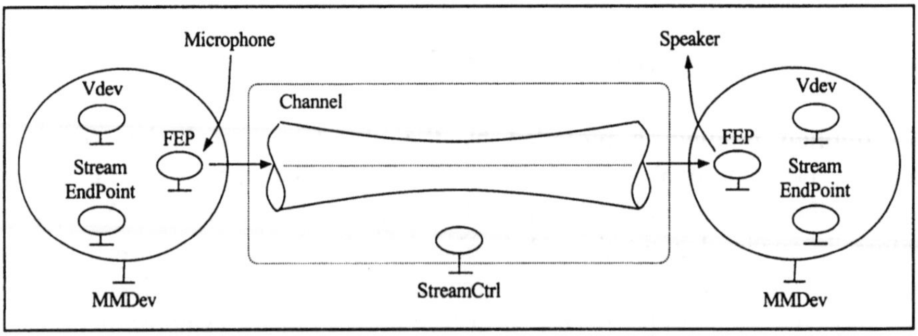

Figure 5. A basic audio stream configuration

As stated before, our DPE implementation provides mobility of clusters through deactivation and subsequent reactivation on a different capsule. One important feature of our DPE is the incorporation of mobility to channels. When a cluster migrates, all channel endpoints are equally deactivated and reactivated at the new cluster destination. This feature allows a TINA session to be suspended at one site and resumed on a different site.

In order to incorporate mobility to channels, we specialized our AVStreams implementation, mainly the stream controller (`StreamCtrl`) and the multimedia device (`MMDevice`) objects. The idea is to incorporate mobility functions to stream controllers and checkpointing/recovering functions to multimedia devices. Such functions are absent in the current OMG AVStreams specification. The extended `MMDevice` IDL interface is given below:

```
interface AppMMDevice : AVStreams::MMDevice {
  long checkpoint(in string template);
  long recover(in string template);
  long Delete();  };
```

This extended interface just turns an AVStreams multimedia device object into an engineering computational object (eCO) by incorporating the `checkpoint()`, `recover()` and `Delete()` operations.

The extended `StreamCtrl` IDL interface is presented below:

```
interface ChannelCtrl : AVStreams::StreamCtrl {
 long configureChannel (in AVStreams::streamQos the_qos,
      in AVStreams::flowSpec the_spec);
 long configureEndPoints (in eCOManager a_eco, in AppMMDevice
      a_party, in eCOManager b_eco, in AppMMDevice b_party);
 long changeEndPoint(in eCOManager eco, in AppMMDevice party);
 long deactivateChannel();
 long reactivateChannel();  };
```

This extended interface allows the configuration (`configureChannel()`) of the channel by stating the quality of service parameters (e.g., network bandwidth and maximum delay) and the specification of the flows (e.g., PCM audio and MPEG video). `configureEndpoints()` establishes the binding between two multimedia devices, assigning them to their respective eCOs. `changeEndPoint()` allows the re-assignment of a multimedia device to a new eCO. This operation is performed when an eCO attached to a channel is reactivated after migration. `deactivateChannel()` and `reactivateChannel()` operations are performed during the migration process.

Two more operations are related to channel creation and destruction and are part of the node interface (as specified by RM-ODP):

```
interface NodeManager {
  long createChannel
    (in eCOManager a_eco, in AppMMDevice a_party,
     in eCOManager b_eco, in AppMMDevice b_party,
     in AVStreams::streamQos the_qos,
     in AVStreams::flowSpec the_spec,in ChannelCtrl the_ctrl);
  long destroyChannel(in ChannelCtrl the_ctrl);  };
```

TINA applications call `createChannel()` in order to establish a stream binding between two eCOs. The application passes the references of the multimedia devices (a_party and b_party) and the eCO managers assigned to the eCOs to be bound. QoS parameters and flow types (the_qos and the_spec) may also be specified (null parameters means "no QoS specified" and "all flows", respectively). The last parameter is a reference of a channel controller that will manage the channel. Each channel type supported must implement a channel controller of that type. Our DPE has two channel controllers implemented: one for PCM audio and one for CellB video. `createChannel()` returns a flag indicating success or failure. Upon success, the application uses the channel control object to start, pause and destroy the channel.

When a cluster migrates, the DPE takes the following actions with respect to channels. Each eCO manager maintains a list of channel controllers managing the channels connected to the object. When the object is deactivated, the eCO manager calls the `deactivate()` operation on the channel controller in order to de-allocate the resources assigned to the channel. When the object is reactivated in the destination capsule, the eCO manager re-establishes the channel by calling `changeEndPoint()` and `reactivateChannel()` on the channel controller. Notice that the channel controller is not deactivated during migration of channel endpoints.

In case of point-to-point channels both channel endpoints are destroyed and reconstructed during migration. For point-to-multipoint channels, only the migrating endpoint is destroyed and reconstructed during migration.

5. IMPLEMENTATION ISSUES

The TINA DPE reported in this paper was implemented on the following computing infrastructure. Processing nodes consist of a set of Sun Sparc 5 and Ultra workstations with OC-3 ATM network interface card (155 Mbits/s) connected through an ATM switch from Xylan Corporation. The Solaris 2.6 operating system from Sun Soft runs on each processor. Orbix 2.3 [11], a CORBA 2.0-compliant platform from Iona Technologies, and ObjectStore 5.0 [12], a distributed object-oriented database from Object

Design, are the major software products employed in this project. Orbix provides the major CORBA components (IDL compiler, ORB, implementation and interface repositories and locator), while ObjectStore provides persistence functions necessary for checkpointing and recovering of objects, clusters and channels.

At each node runs a CORBA server (servant) implementing the `NodeManager` interface. Capsules are Solaris processes with a servant implementing the `CapsuleManager` interface. This servant executes on an exclusive thread. Cluster and eCO managers are also servants running on exclusive threads. Application servants are distributed through any number of threads. All of these threads execute within the capsule's address space. This implementation follows exactly the RM-ODP where capsules provide the resources for clusters and objects.

In order to illustrate the DPE support for migration, let us consider a "desktop telephone" service[3] (Fig.6). Before the establishment of the audio stream supporting the conversation, two capsules are created on each participating node. The application at the caller side calls `createChannel()` on its node interface in order to establish an audio channel with two flows connecting the multimedia devices (a pair of microphone and speaker at each side). These devices derive from the interface `AppMMDevice`. An object that implements the interface `audioChannelCtrl`, a specialization of the interface `ChannelCtrl` for audio streams, controls the audio channel.

Suppose that one party decides to move to a different node. The application can then build an empty capsule at this node and call `migrate()` on the actual capsule passing the new capsule as target. After the cluster migration completes, the channel is re-established and the application resumes. It is important to notice that TINA applications hide the DPE functions from the final users.

6. CONCLUDING REMARKS

This paper presented a CORBA-based implementation of a major component of the TINA architecture, the Distributed Processing Environment (DPE). The implementation has the following highlights: support to deployment and management of nodes, capsules, clusters and objects; support to stream binding, allowing application objects to exchange media flows; support to migration, allowing clusters to migrate from one capsule to another; multi-threaded implementation, allowing different levels of parallelism among the objects.

[3] The telephone set is a desktop computer and the telephone networks is a data network.

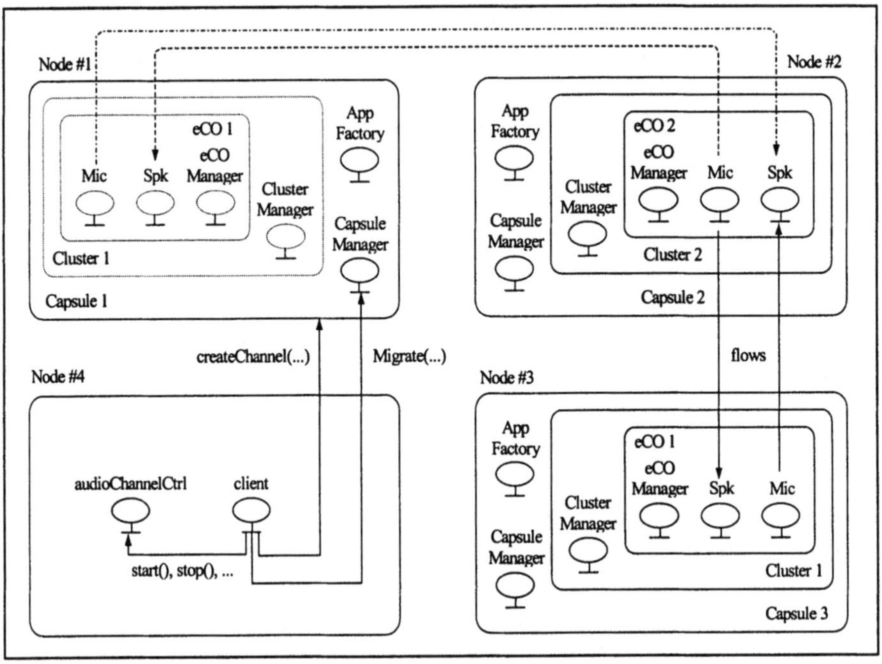

Figure 6. Desktop telephone application.

Among the applications of interest, the offering of telecommunication services over the Internet is receiving special attention. The increasing of speed in both on access networks and backbones is making the Internet the most promising infrastructure for new telecommunication services. Since the TCP/IP protocol stack provides only connectivity services, the integration of TINA and Internet can turn the Internet in a truly multi-service global network. Another point of interest is to employ some CORBA 3 [13] features, mainly, objects supporting multiple interfaces[4] and quality of service control. These features are not available in the current CORBA development platforms.

ACKNOWLEDGMENTS

This research is being supported by the following Brazilian funding agencies: CAPES which provides MSc and PhD scholarships to the first and second author, CNPq (grant 300723/93-8) and FINEP (grant 1588/96).

[4] Multiple interfaces allows a direct mapping between TINA and CORBA objects.

REFERENCES

[1] R.W. Lucky. "New Communications Services – What Does Society Want?", Proceedings of the IEEE, 85(10), October 1997.

[2] Fabrice Dupuy, Gunnar Nilson, and Yuji Inoue. "TINA Consortium: Toward Networking Telecommunications Information Services", IEEE Communications Magazine, 33(11): 78-83, November 1995.

[3] Object Management Group. "CORBA/IIOP 2.3.1 Specification", Technical Report formal/99-10-07, 1999. URL at http://www.omg.org.

[4] ISO/IEC 10746 / ITU-T Draft Rec. X.901, X.902 & X903. "ODP Reference Model Part 1, Part 2 & Part 3", Technical Report, June 1995.

[5] H. Berndt et al. "The TINA Book". Prentice Hall Europe, 1999.

[6] TINA-C. "Engineering Modelling Concepts - DPE Architecture", Technical Report, TINA Consortium, December 1994. URL at http://www.tinac.com/.

[7] S.Vinoski. "CORBA: Integrating Diverse Applications Within Distributed Heterogeneous Environments", IEEE Communications Magazine, 14, February 1997.

[8] L.F.Faina, E.J. Oliveira, R.C.M. Prado and E. Cardozo. "Developing Multimedia Applications with the OMG Streaming Framework". In Proc. of the ICC'99 - MAST'99 (Mini Conference on Multimedia Applications, Services and Technologies), Vancouver, Canada, June 1999. URL at ftp://ftp.dca.fee.unicamp.br/pub/docs/eleri/Mast99.pdf

[9] A.S. Pinto, E.J. Oliveira, L.F. Faina and E.Cardozo. "TINA-based Environment for Mobile Multimedia Services". In Proc. of the TINA'99, Hawaii, USA, April 1999.

[10] Object Management Group "CORBAtelecoms: Telecommunications Domain Specification", Technical Report formal/98-07-12, OMG, October 1997. URL at http://www.omg.org/.

[11] IONATechnologies Ltd., "Orbix 2.3c MT - Programming Guide", November 1997.

[12] Object Design. "ObjectStore - C++ API User Guide - Release 4", June 1995.

[13] Jon Siegel, "CORBA 3 Fundamentals and Programming", John Wiley & Sons, second edition, 2000.

Chapter 13

Development of Parlay-based Services Using UML and SDL

Markos Koltsidas, Ognjen Prnjat, Lionel Sacks
Departement of Electronic and Electrical Engineering,
University College London, Torrington Place, London WC1E 7JE, U.K.
Phone: +44 20 7679 3198   Fax: +44 20 7387 4350
Email: {oprnjat | lsacks}@ee.ucl.ac.uk

Key words: Service development approach, Parlay API, SDL, UML.

Abstract: Business and residential demands for sophisticated multimedia services are constantly on the increase, while the rapid introduction of new services and the user-tailored customization are emerging as the key requirements. A number of initiatives in the area of programmable network interfaces are paving the way for the future telecoms environment where new, customized services will be developed more quickly and efficiently by the third party providers and end-users, and deployed over the operators' network and support systems. Parlay API is one of the most notable examples of an open network programming interface, its development being carried out by the key players in the industry. Parlay specifications define a number of interfaces available to service developers, and some basic guidelines on the way of accessing these interfaces: however, no clear development approach is defined to guide the developers. Here we present and assess an approach for specification and design of the Parlay-based 3rd party applications, based on UML and SDL. We illustrate the approach with the case study of the "Conference between three parties" application.

1. INTRODUCTION

Nowadays the network operator controls all the telecoms network capabilities and is wholly responsible for the design, development and operation of all services. The existing environment is effectively made up of three layers: the bottom layer representing different network technologies, the middle layer consisting of service components such as routing, configuration *etc.*, and the top layer featuring various applications. This architecture is sufficient for provision of services that satisfy common customer needs but the customized end-user services are not quickly and easily deployable. Overall time to market for new services is high since only the operator's service development team is involved in application development. High service deployment time consequently impacts the customer base and the operator's revenue: both residential and business customers are increasingly asking for new, sophisticated services that can be tailored to their demands. On the other hand, deregulation activities, such as the Open Network Provisioning in Europe, are prompting the established operators to allow the other licensed operators and service providers the access to their networks and support systems. Thus, operators are looking into the possibility of opening up their network and control systems at the considerably lower level of access then was available in the past. This would allow the 3rd party application developers to specify, design and deploy new services over the operator's infrastructure, using the network application programming interfaces (open network APIs). One of the main initiatives in this field is lead by the Parlay group, which is developing the Parlay API.

2. PARLAY API

Parlay API is an open technology-independent interface to generic network service components in the operator's domain: it is located between the application and the service component layers (*Figure 1*). Its goal is to allow applications to be developed and deployed by end users / value-added service providers outside the network domain. Parlay API has a revolutionary potential within the telecoms industry: software developers will have the ability to design and offer sophisticated services, by gaining access to operators' generic service components and networks.

Parlay API is object-oriented and consists of two categories of interfaces: service and framework interfaces. Service interfaces allow the application to access a wide range of network capabilities. Parlay phase 1 specification provides an interface to call control and messaging, while phase 2 gives access to wireless and IP services. Framework interfaces provide additional management support for service interfaces, such as discovery, notification, authentication, *etc.*

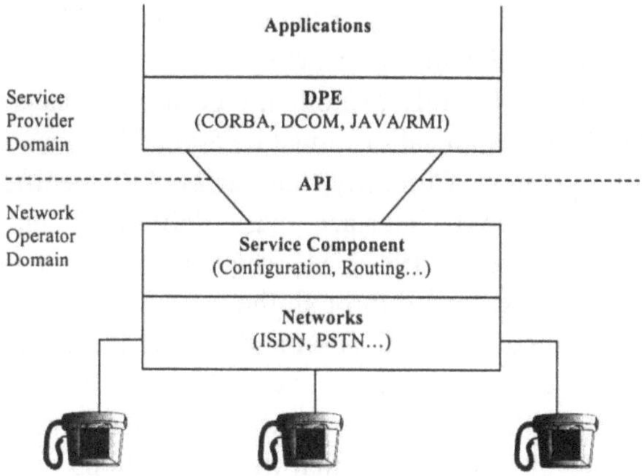

Figure 1. Future telecommunications network and service environment: the Parlay API

Parlay API Specification [Parlay99] consists of a set of documents. **Core Specification** defines the interfaces, methods, parameters and a few state models of the API. **Data Definitions** document specifies the data definitions. **Sequence Diagrams** document specifies the basic services, through the use of sequence diagrams, that can be derived from the API. **Class Diagrams** document describes the class diagrams used by the API specification and recommends how these should be presented. **MIDL Specification** document describes Microsoft (SOM) Interface Definition Language for the Parlay API. **OMG-IDL Specification** document describes OMG Interface Definition Language for the API. Core Parlay API interface specifications are technology-independent (specified in Unified Modeling Language [UML]) and thus potentially support a wide range of distribution middleware such as CORBA, DCOM and JAVA/RMI. Combining Core Specifications with the MIDL specification will support DCOM, while combining the Core Specifications and the OMG-IDL specification will support CORBA.

Introducing Parlay API brings many benefits to all the players in the market. For the operators, the API can become the key factor for the progress of the market through the increasing demand for new services, which can be created and operated in short time by collaboration with service providers. Service providers have the opportunity to create new services or make improvements to existing ones in relation to the API capabilities. As a result, the service reaches the market faster. Application developers also benefit from the Parlay API: applications that were either previously not technically feasible (network operator could not provide the necessary functionality) and/or not financially acceptable (high cost of development) can now be developed. End users have the ability to integrate new application solutions with other software. Competition between developers results in the reduction of the price that is offered from an open market for customer applications [Parlay99].

For these benefits to be fully exploited, a clear and concise development approach for third party applications must be agreed between the developers and the network operators. Currently, the Parlay API specifications consist solely of the simple UML interface class definitions and a few basic examples of the interface use (via UML sequence diagrams). Behaviour of the interfaces is poorly specified through the use of a few simple UML state diagrams. The exchange of specification and design documents and a certain level of collaborative development between the application developers and network operators will be inevitable in the future. Moreover, operators will require a certain level of assurance that the applications developed outside their own domain will not in any way harm the correct and proper functioning of their service components and the networks. The operators will need a proof that the integrity of their domain will not be compromised. Integrity is defined as the ability of the system/network to retain its attributes in terms of performance and functionality [Ward95].

Thus, a development approach that will ensure both clear and concise mutual understanding, between the operators and developers, of the service application functionality, as well as allow a certain level of application integrity analysis, and verification and validation, is needed.

3. SERVICE DEVELOPMENT APPROACH

First step in the application development requires thorough understanding of the Parlay specification documents discussed above. Having consulted these, and prepared the textual description of the target application, the developer can use the following development phases:

- Application requirements analysis and high-level design using **Unified Modeling Language** (UML) notation [UML].
- Detailed application design using the **Specification and Description Language** (SDL) [Z.100].
- Application **integrity analysis** based on specification/design documents; **simulation and validation** using widely available SDL tools.

Application requirements analysis and high-level design using UML. UML is a third-generation object-oriented language, rapidly becoming a software industry standard. It is more expressive yet cleaner and more uniform than OMT [Rumb91] or other established visual modeling languages. UML is used to specify, construct and document structure and behaviour of a software system. It consists of a number of basic diagrams and core concepts, so the users can easily learn and use it, but also extend it. Parlay API specification is based on UML: it uses class diagrams for interface descriptions, sequence diagrams for description of simple example applications, and state diagrams for depicting some basic behaviour. Here we advocate a more thorough use of UML, where the development would start by specifying the high-level application functionality through the use-case diagrams. Then, the classes needed to support this functionality would be chosen from the Parlay Core Specification document, and their relationships would be depicted through class diagrams, while class internal behaviour would be depicted through state diagrams. Interaction between the objects implementing the interfaces is depicted through collaboration and sequence diagrams, showing how objects collaborate to perform a use-case.

Detailed application design using SDL.

SDL is a formal description technique based on the concept of extended Finite State Machines. This sophisticated modeling and design technique allows complex behaviour to be expressed through asynchronous communication between a number of objects containing processes. UML, although clearer, cannot fully describe this complex behaviour. SDL is the most widely accepted formal language in the telecoms industry [Loge99], and is supported by a number of software tools. Detailed SDL design of the Parlay-based application is easily constructed from UML analysis and high-level design documents: Parlay interface objects map to the SDL objects/processes; behaviour of the processes is derived from UML state diagrams, and inter-process interaction from sequence and collaboration UML diagrams. This yields a detailed design of the Parlay-based application and the supporting objects, a design capable of depicting the complex behaviour by exploiting the powerful features of SDL. SDL design is easily transferable into executable code through the use of widely available tools.

Application simulation, validation and integrity analysis.

SDL is supported by a number of tools, which provide the means of simulating the design so as to produce Message Sequence Charts (MSCs). MSCs visually depict the behaviour of the detailed design, as a response to the external stimuli. MSCs produced in this manner can be compared to the analysis UML sequence diagrams so as to verify the conformance of the detailed design to the requirements specification. Moreover, the availability of the detailed behavioural design in SDL allows both the network operators and service providers to conduct the application and supporting Parlay objects' integrity analysis [Prnj99][Prnj00], in terms of a number of factors. Using SDL validations tools, the correct behaviour of the SDL design can be conducted by exploring the state space of the application. Reachability analysis can be performed, and livelocks and deadlocks can be detected. Moreover, conducting the service design in SDL can greatly help in feature interaction detection prior to service launch. Network operators having access to a number of SDL designs could compare and simulate them so as detect feature interactions [Came93] that might arise due to the unexpected interactions between a number of correctly designed applications.

4. CASE STUDY

This section illustrates our development approach using the "Conference between three parties" example, which provides an audio conference between three users. The application allows user A, who is in full control, to establish through the Parlay API a conference call with user B, and then, at any point of the call, user A can add user C to the existing conference.

4.1 UML requirements analysis and high-level design

This phase involves analyzing the Parlay Core Specification and elaborating on it through a number of UML diagrams. Application high-level requirements and functionality are first captured via a **use-case diagram**. The diagram shows the structure of the system as a graph of actors, a set of use-cases enclosed by the system boundary, communication associations between the actors and the use-cases, and generalizations among use-cases.

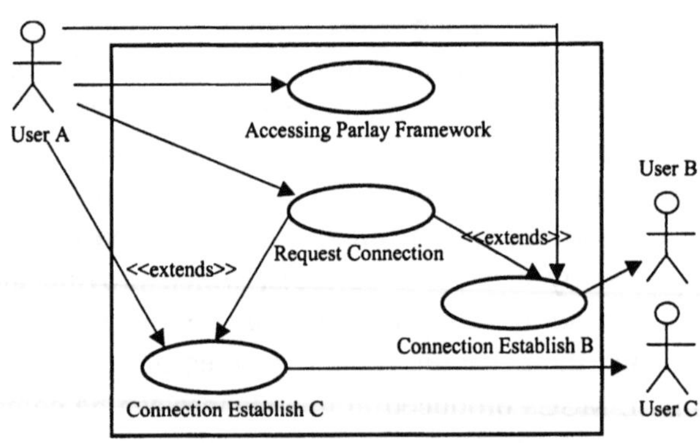

Figure 2. "Conference between three parties" use-case diagram

Figure 2 depicts the use-case diagram for the example application. First, user A accesses the Parlay API (the framework interface). Then, A is using the application in order to request connection with user B and then with user C. Use-case "Request Connection" extends in "Connection Establish B" and "Connection Establish C" use-cases. An "extends" relationship from use-case "Request Connection" to these use-cases indicates that an instance of these use-cases may include the behaviour specified by the parent use-case. When A has established a connection with B, he/she requests and establishes connection with C, thus creating a three-party conference call.

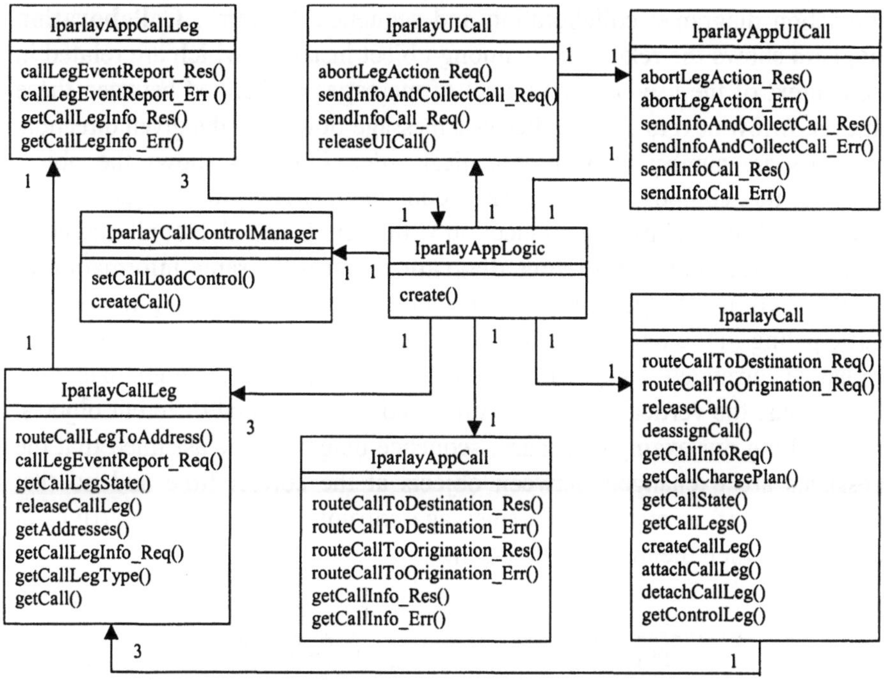

Figure 3. "Conference between three parties" class diagram

Having captured the application requirements, we must consult the Parlay API Core Specification document for the choice of interface classes needed to support the use-cases. A number of classes were chosen to support our application. *IparlayAppCall* handles call request responses and state reports. *IparlayCallControlManager* creates call objects. *IparlayCall* provides a structure to allow complex call behaviour to be used. *IparlayUICall* provides functions to send information to, or gather information from, the user to which a call leg is connected. *IparlayAppUICall* handles call user interaction request responses and reports. *IparlayCallLeg* tracks its own state and allows charging summaries to be accessed. *IparlayAppCallLeg* handles responses and errors associated with requests on the corresponding callLeg. These interface classes, the class operations, and the relationships between the classes, are shown on the **class diagram** on *Figure 3*. There are three instances of IparlayCallLeg and IparlayAppCallLeg, corresponding to the three users involved in the conference call. The arrows in the diagrammatic representation indicate the way across which the communication flows: a line without arrows indicates that the communication takes place both ways.

The next step is to depict how the instances of these classes interact in order to fulfil a use-case. This can be done using two types of UML interaction diagrams: collaboration and sequence diagrams. **Collaboration diagram** shows the relationship among object instances which are related in the context of the use-case, and a set of messages exchanged between these objects. Arrow next to a line indicates a message flowing in the given direction. Messages are numbered in chronological order. *Figure 4* shows the object instances that compose our example model and messages that are exchanged when user A is establishing connection with users B and C. Not all messages are shown, for clarity. **Sequence diagram** captures the same information as the collaboration diagram, but it also includes time, by showing the objects participating in the interaction by their "lifelines" and the messages that they exchange arranged in a time sequence. Sequence diagram has two dimensions: the vertical represents time, and the horizontal different objects. In modelling our example scenario sequence diagrams, we assumed that the messages are exchanged between objects at the correct time and no that errors occur. If an error occurs (message corrupt, message never arrives at destination) then the message is re-sent.

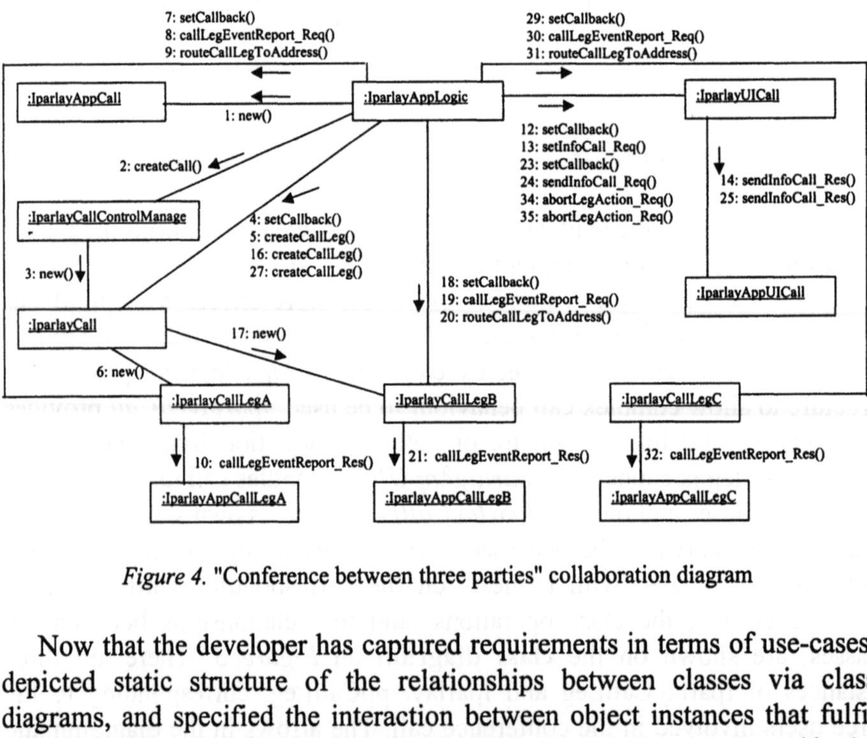

Figure 4. "Conference between three parties" collaboration diagram

Now that the developer has captured requirements in terms of use-cases, depicted static structure of the relationships between classes via class diagrams, and specified the interaction between object instances that fulfil certain functionality, the last step in the high level design is to specify the internal behaviour of the classes by developing a state diagram for every class chosen for the target application. **State diagram** shows the sequence of states that an object traverses during its lifetime in response to received stimuli. It thus represents a state machine: a graph of states and transitions that describes the response of an object to the outside input. A state is a condition in the life of an object during which it satisfies some condition, performs some action, or waits for some event [Fowl97]. *Figure 5* illustrates one of the state diagrams for our example application. This state machine for the IparlayCallLeg interface shows all the possible states of the object as it receives different messages, while the arrows represent state transitions.

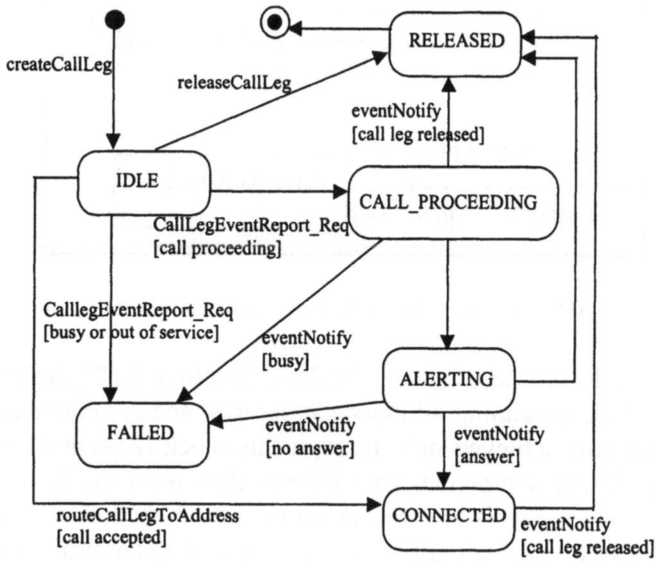

Figure 5. IparlayCallLeg state diagram

4.2 Detailed application design using SDL

On the basis of the UML requirements analysis and high-level design, the detailed SDL design can be developed. SDL model consists of concurrent objects, described by processes, which define the control structure in terms of states and transitions. Objects are grouped into blocks, and communicate via signals, which are sent through channels. The UML to SDL mapping is straightforward: UML object instances map to SDL objects, each described by a process diagram derived from the UML state diagram. Signals are derived from UML collaboration and sequence diagrams.

Starting from the high-level UML design, our example application is designed in SDL. The SDL system design consists of a number of entities. The *Parlay* **system diagram** defines the high-level architecture of the system (*Figure 6*). The *ParlayAPIObjects* **block** is effectively the container of **objects. Processes** (finite state machines extended with data) then describe the objects in terms of states and transitions. **Signal lists** define groups of signals that are exchanged between the objects/processes. The objects, which make up the SDL system, communicate through **channels**.

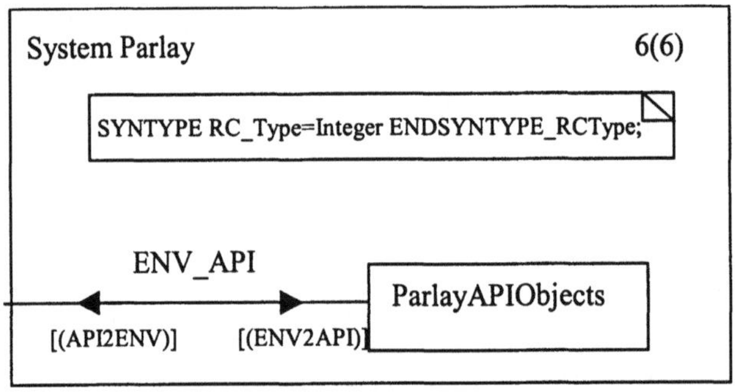

Figure 6. "Conference between three parties" SDL system diagram

Our design, developed using the Telelogic Tau tool SDT [Tau3.5], uses a single SDL block containing all objects/processes, and communication with the environment is achieved only through this block (*Figure 6*). By using only one block and one signal list on the system level the developer can easily add or delete signals and thus avoid signal duplication, that can occur when a number of blocks are used, and that can lead to an error. The objects contained in the SDL block are mapped directly from UML object instances: *IparlayAppCall*, *IparlayCallControlManager*, *IparlayCall*, *IparlayUICall*, *IparlayAppUICall*, *IparlayCallLegA*, *IparlayCallLegB*, *IparlayCallLegC*, *IparlayAppCallLegA*, *IparlayAppCallLegB*, *IparlayAppCallLegC*. Additionally, an *Error_Log* procedure was created to handle the unexpected signals: if a process receives an unexpected signal, with the help of the exception-handling Error_Log procedure it stays in the same state. All processes have been designed so as to react to a correct incoming signal by moving to a relevant state. The sound behaviour of the SDL design depends on the correct design of the processes and their states. Enriching the SDL model can be achieved through the consideration of new scenarios and creation of new states for the processes to enable them to handle various incoming signals. As a result, the SDL model becomes behaviorally rich and is capable of accommodating a wide range of scenarios.

4.3 Simulation, validation and integrity analysis

The final phase in the development approach is the detailed SDL design simulation, validation and integrity analysis. Simulation means executing the system under developer's control: signals can be sent to the system in order to examine its behaviour. Through simulation, developer creates Message Sequence Chart (MSC) diagrams, which depict the details of the interaction between the processes and the states that the processes go through.

Figure 7 illustrates an MSC diagram of our example application design simulation, representing the End_of_Conference. This diagram is only a part of the overall MSC diagram that was created with the help of the Telelogic Tau *simulator tool*. This MSC represents the exchange of signals during the end of conference between users A, B and C. In this case user A decides to first interrupt the conversation with user C and then with user B.

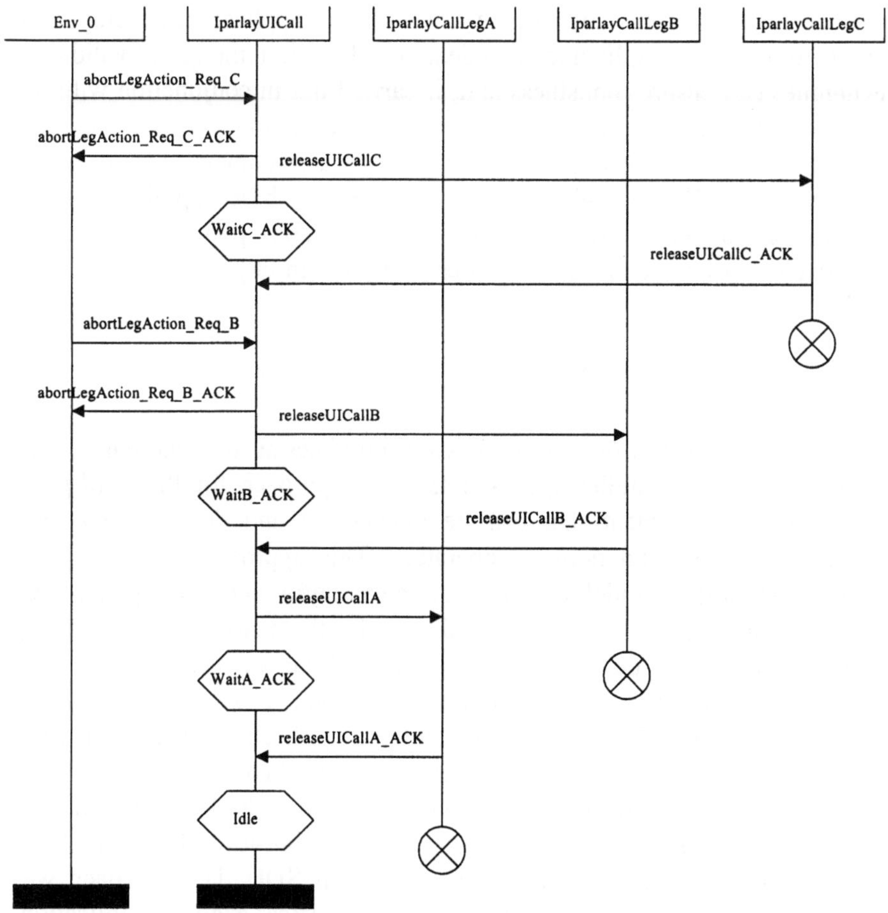

Figure 7. Message Sequence Chart depicting the end of conference

MSCs produced in this fashion are then compared with the initial requirements capture UML sequence diagrams. If the diagrams match for all the scenarios, the developer can have confidence that the detailed design conforms to the specification. In the case of our example application, the diagrams are nearly identical: discrepancies occur in the cases where the important signals in the SDL design were made to be acknowledged by a special message, a feature that was not included in the requirements analysis.

Moreover, the SDL tool used for our design could perform an in-depth validation of the application, with the SDL *validator tool*. This tool allows the exploration of the overall state space of the SDL system and is capable of detecting a range of run-time errors, by performing reachability analysis and livelock and deadlock detection. We did not use this feature, since it required the inclusion of timers for every signal exchange, while our application was not designed as a real-time one. However, this is a powerful feature, which should be fully exploited by application developers and network operators so as to detect and resolve any integrity risks during application development. Apart from performing livelock and deadlock detection, the use of validation techniques can ensure robustness and, if carried out in conjunction with the integrity analysis of other potentially interacting applications, can also resolve feature interaction problems in the operator domain. This does not guarantee the fully integral operation of the Parlay-based applications and the network operators' support systems, but it does partially ensure the required level of some integrity features [Prnj99][Prnj00].

5. DISCUSSION AND CONCLUSIONS

Future prospects of the use of network interfaces such as the Parlay API raise the issue of establishing a standardized approach for the third-party application development that can be understood and deployed by both service developers and network operators. This approach should support clear development guidelines and use meaningful and widely accepted notation schemes, and also give a solid basis for the analysis of the integrity features of the newly developed applications and the way these interact with the operator's control systems and the network. Here we presented a development approach for the Parlay-centered applications, based on UML and SDL. Development phases were presented, leading the developer from the basic application textual descriptions and Parlay specification documents, through the requirements capture and high-level design in UML, to the detailed and behaviourally rich design in SDL. The approach was illustrated using the "Conference between three parties" example application.

The main advantages of this approach are clear and straightforward mappings from the Parlay specifications to UML high-level design, and the UML high-level design to the detailed SDL design. Thus, a developer has a clear guideline how to achieve final service design starting from basic Parlay interface specifications. UML and SDL are widely used languages both in telecoms and IT industries: UML is a de-facto standard, while SDL is increasingly gaining wider acceptance [Loge99]. Incorporating these languages in the approach could help collaborative service development between third-party developers or between developers and network operators themselves. Moreover, UML and SDL can be seen as documentation tools aimed to help the service maintenance: lack of clear service documentation in the operator's domain is more frequent than believed [Ward95].

Another important consideration is that the service modeling techniques used should facilitate close analysis of the design to ensure, at early stages, that the system is complete in meeting its requirements and consistent in its operation. This should yield robust, well-engineered products. Our approach, by providing rich SDL behaviour models, allows the developers and network operators to analyze, validate and test the new applications with respect to their integrity features, with the help of widely available SDL support tools. Service verification and validation can be performed, and deadlock, livelock and feature interaction detection can be conducted prior to service launch. Service validation and integrity analysis are seen to be of paramount importance to the network operators, who will require strong assurance from the third party developers that the applications deployed will not harm their network and supporting systems. The benefits of the proposed approach will have to be traded off against one clear drawback: the SDL development might be time consuming and thus slow down the time to market which can be seen as of significant importance in the future open network scenarios.

REFERENCES

[Came93] E. J. Cameron and H. Velthuijsen, "Feature Interactions in Telecommunications Systems", IEEE Communications Magazine, August 1993, pp. 18-23.

[Fowl97] M. Fowler, S. Kendall, "UML Distilled", Addison-Wesley, 1997.

[Loge99] X. Logean et. al., "On Applying Formal Techniques to the Development of Hybrid Services", IEEE Comms Magazine, Vol. 37, No. 7, July 1999, pp. 132-138.

[Parlay99] Parlay Group, "Parlay API Specification 1.1", Issue 1.1, 7th May 1999.

[Prnj00] O. Prnjat, L. Sacks, "Inter-domain Integrity Management for Programmable Network Interfaces", Management of Multimedia Networks and Services' 2000, September 2000.

[Prnj99] O. Prnjat, L. Sacks, "Integrity Methodology for Interoperable Environments", IEEE Comms Magazine, Vol. 37, No. 5, May 1999, pp. 126-139.

[Rumb91] J. Rumbaugh et. al., "Object-Oriented Modeling and Design", Prentice-Hall, 1991.

[Tau3.5] Telelogic Tau 3.5, SDT Tool, http://www.telelogic.se

[UML] Rational Software, "UML Semantics", Version 1.1, 1st September 1997; "UML Notation Guide", Version 1.1, 1st September 1997.
[Ward95] K. Ward: "Impact of Network Interconnection on Network Integrity", British Telecommunications Engineering, Vol. 13, January 1995, pp. 296-303.
[Z.100] ITU Rec. Z.100, "Specification and Description Language (SDL)", 1993.
[Z.120] ITU Rec. Z.120, "Message Sequence Charts (MSCs)", 1994.

PART FIVE

Policies and Service Management

Chapter 14

Supporting Service Quality Assurance via Trouble Management

Dr R. Sinnott, T. Gringel, Dr M. Tschichholz, W. Vortisch
GMD Fokus, Kaiserin-Augusta-Allee 31, 10589 Berlin, Germany

Key words: Trouble Management, Trouble Ticketing, Service Level Agreements, Service Quality Assurance, CORBA, TINA.

Abstract: The open service market encourages competition between service providers. To attract and keep customers, service providers require – amongst other things - better tools and techniques to increase their competitiveness. In this paper we address one area for tool support: namely, tools for the support of service quality assurance, i.e. so that checks can be made to ensure that services (and the networks they operate over) fulfil the expectations of customers who have subscribed to them. To demonstrate this, we show how trouble management techniques can be applied to develop generic and reusable components. The test-bed for this work is based on a TINA platform Y.TSP that has been extended with a trouble management component. We show how this trouble management component can be used to support service quality assurance via two application cases studies.

1. INTRODUCTION

The deregulation and subsequent opening-up of the telecommunications market offers many potential advantages to service providers and customers. Service providers now have the opportunity to compete in new markets, whilst customers are offered a wider choice of service providers and hence the services they offer. With this in mind, the need for service providers to be more competitive is paramount. This competitiveness can take many forms. It might be through the portfolio of services offered – including add-on services; through the costs associated with the access and usage of those services; through the quality of the services on offer etc. In this paper we address one area of competition: *service quality assurance*. Put simply, this is the ability to check that services that customers subscribe to fulfil their expectations. When this is not the case, e.g. due to problems (troubles) with the services, then appropriate actions are taken to resolve the problems and potentially, to offer some form of compensation to the affected customers. It is precisely this level of flexibility, i.e. offering compensation for problems that have occurred, that help to distinguish – and hence make more competitive! – different service providers.

As well as direct benefits to the customers through making the service providers more competitive, components that support the ability to monitor and potentially resolve troubles can also be applied to the direct financial advantage of the service provider. Thus for example, these same components can be applied to try to minimise any discounts that might have to be given to customers affected by troubles.

Troubles or failures more generally themselves can take many forms. It might be the case that complete hardware and software failures occur, in which case automated trouble management requires some form of backup, e.g. through contact phone numbers or addresses of people capable of resolving the problem. A less drastic form of failure might be through a certain feature of a service not working properly, e.g. the throughput of a service is not high enough. We shall see how trouble management techniques can be applied to cover a myriad of different failures affecting different systems using different technologies. We note that this incorporates failures that occur at different system levels, e.g. at the service level and network levels. What is crucial is that the information related to the troubles can be passed to the appropriate management systems and possibly the people or administrators responsible for resolving the problem.

It should be pointed out that in the open service market, integrating different systems offering potentially disparate trouble handling or fault management systems developed with different business processes in mind is a non-trivial matter. It is often the case that the work on integrating trouble management systems initially requires, work on integration of business processes. In this paper we focus predominantly on CORBA [2] based technologies applied to trouble management and do not address integration of business processes. Work in this area is documented in [6].

The rest of the paper is structured as follows. Section 2 gives an overview of trouble management and presents the main ideas behind trouble tickets and their application. Section 3 provides an overview of the architecture of the systems used to demonstrate trouble management. Section 4 focuses upon the service based trouble management components and discusses their design. Section 5 illustrates the application of the trouble management components via two cases studies: a music on demand service (MusicShop) and a image distribution service developed as part of a Dutch Auction flower service. Finally we draw some conclusions on the work and identify future extensions.

2. OVERVIEW OF TROUBLE MANAGEMENT

Dealing with troubles that can potentially occur in different administrative and technological domains is a complex activity, especially for open systems, i.e. not simple interworking between a closed set of software elements. Ensuring that the necessary information is transferred to the appropriate components, or if necessary, to the people in charge of finding and identifying the trouble, is a complex activity. This is especially so where numerous different types of failure can occur with numerous causes and manifestations. To help provide a level of structuring to this open-ended problem, various standards [18], initiatives [15-17] and research projects [3,4,9] have taken place. A common approach taken by these has been through the notion of *trouble tickets*. Loosely, a trouble ticket is something that allows for the lifecycle of a trouble or failure to be monitored and acts as a basis for transferring the appropriate trouble information between different management systems. Figure 1 highlights the state diagram of a trouble ticket.

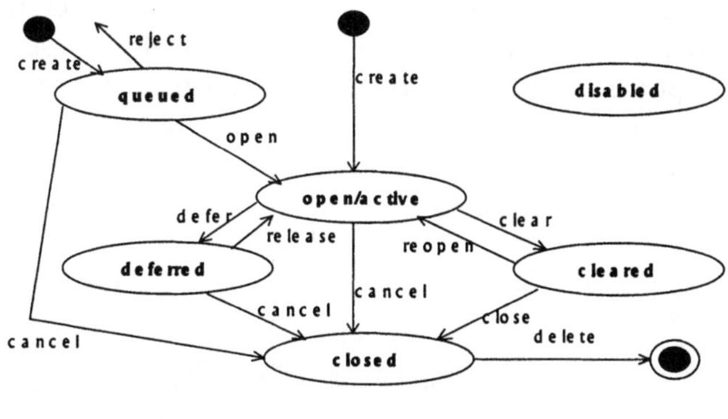

Figure 1. State Diagram for a Trouble Ticket

Trouble tickets can be created in state queued (where the trouble monitoring and resolution has not yet begun) or state open/active in which case the trouble is being worked on. A trouble ticket might be rejected if no such trouble exists. Once a trouble has been resolved, i.e. the failure has been resolved by a suitable person or fault management system, the trouble ticket moves to state cleared. It might be the case at this stage that the affected parties, e.g. the customers are requested to verify that the service or network is now functioning correctly – or at least that the trouble previously reported has been resolved satisfactorily. If this is the case then the trouble ticket moves to state closed where the information associated with the lifecycle of that trouble can be written to a log. This might then be used for administrative purposes, e.g. to perform statistics on the troubles that have occurred and their causes etc.

Trouble tickets can also be disabled or deferred depending upon whether the information related to the trouble ticket cannot be updated for a particular reason, or the trouble itself cannot be dealt with at a given time.

It should be emphasised that disparate parties can be the instigators for the creation of a trouble ticket, e.g. a customer might complain about a service exhibiting a particular failure. Alternatively, it might be the service itself that informs customers about problems it is having – or will potentially have. Another alternative is that it is the network that the service makes use that might be the source of the initial trouble identification. The point is that troubles can occur anywhere and be identified by numerous sources.

The simplistic idea of trouble tickets lend themselves to supporting a generic and open approach for trouble management, i.e. they provide a common core functionality for a plethora of troubles to be realised, be they at the service or network levels. To achieve this, a standardised way to transfer trouble information is necessary. The ITU-T standard X.790 defines various managed object classes which support a broad spectrum of potential failures that might occur with ITU-T applications [18]. This document was used as a basis for the CORBA based SMART Customer to Service Provider Trouble Administration documents [15-17]. Essentially, these documents provide descriptions of the interfaces that should be supported by trouble management systems so that trouble information can be passed around successfully. The following diagram provides a snapshot of the functionality presented in the SMART document.

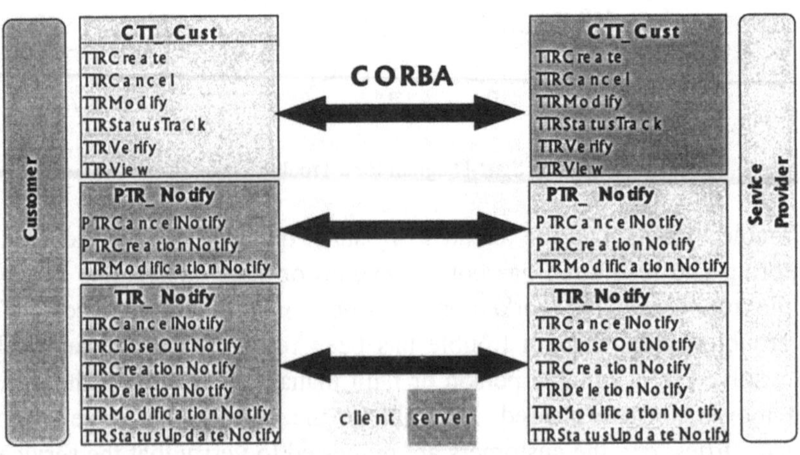

Figure 2. SMART CORBA based Customer to Service Provider Interfaces

These interfaces allow for the sending of the informations related to trouble tickets, e.g. to create trouble tickets (TTRCreate), to update the status of existing trouble tickets (TTRStatusUpdateNotify), to close existing trouble tickets (TTRCloseOutNotify) etc. Since troubles themselves can take a myriad of forms and involve numerous software and hardware components these interfaces typically possess many detailed lists of parameters. Through setting appropriate values for these complex sequences of parameters and providing appropriate information related to the troubles themselves, e.g. their nature and where they are occurring, appropriate trouble management systems can deal with the troubles.

Recent works have shown [1,20] how it is possible to build integrated inter-domain trouble management systems using these technologies. Specifically, the ACTS FlowThru project [5] has shown how gateways, e.g. for TMN-CORBA interworking, and generic trouble ticketing components could be built to support integrated trouble management. We note also that interworking of trouble ticketing systems using TMN technologies was also undertaken in the Eurescom Project P612 [3,4]. Our emphasis here is on one of the CORBA based trouble ticketing components that was developed in the FlowThru project: the TINA Trouble Ticket Reporting System (TTRS). In particular we show how this component has been integrated within the GMD Fokus Y.TSP platform [19] so that it can deal with troubles that might occur with services and networks, and with the consequences of those troubles.

3. TINA ARCHITECTURE AND Y.TSP PLATFORM

The TINA architecture [9] defines a set of concepts, principles, rules and guidelines for constructing, deploying, and operating CORBA based services. The major principles are based on the Reference Model for Open Distributed Processing [8]. The purpose of these principles are to insure interoperability, portability and reusability of software components; independence from specific technologies and to share the burden of creating and managing systems among different business stakeholders, such as consumers, service and connectivity providers [18]. Reference Points are defined to specify conformance requirements for TINA products [13].

TINA provides a set of specifications, e.g. Computing Architecture, Distributed Processing Environment Architecture, Service Architecture and Network Resource Architecture [10-14]. The Service Architecture introduces the underlying concepts and provides information on how telecommunication applications and the components they are built from, have to behave. Central to the Service Architecture is the concept of a *session*. Three sessions are identified: the **access session** which represents mechanisms to support access to services (service sessions) that have been subscribed to; the **service session** which includes functionality to execute and control and manage sessions, i.e. it allows control of the communication session; the **communication session** which controls communication and network resources required to establish end to end connections.

Various components have been identified as being useful in the different sessions. Of relevance here are the access session related subscription and accounting components. The accounting component is used to store the charges/discounts a user incurs when using a given service. This includes the service and network charges. The subscription component is used to store customers subscription information, e.g. the services they have subscribed to and the agreements they have on the usage of those services.

The access session user application (asUap), provider agent (PA), initial agent (IA) and user agent (UA) enable service selection and secure service access in a TINA environment. The relation between these and the service session components is given in Figure 3.

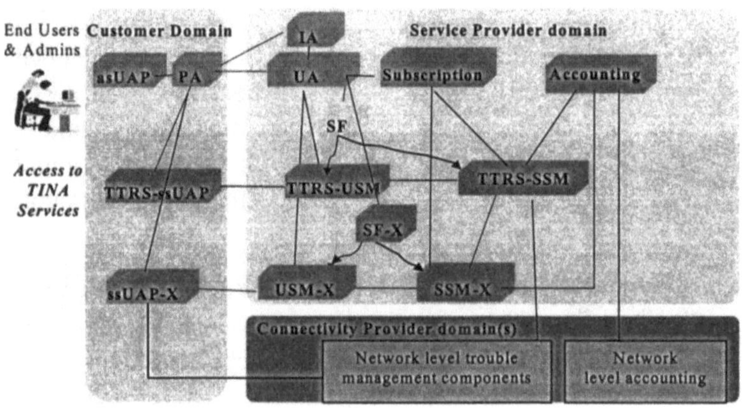

Figure 3. Description of the Y.TSP Components

The service session components typically consist of a service factory (SF) which is used for starting instances of services ofen represented as a service session manager (SSM) used to control the global logic of the service and one or more user service session managers (USM) used to control each users participation (and state) in that service. Users interact with the service itself through some service session specific user application (ssUAP). We note that the Y.TSP platform effectively provides a complete infrastructure in which services - themselves represented as service session components - can be deployed and used with minimal integration effort.

Included in the above diagram are the service session related components for the trouble ticket reporting system (TTRS). This is the component used to facilitate trouble management in the Y.TSP platform. The TTRS service can be subscribed to as with any other service deployed in the platform. Once subscribed to however, this service is started automatically once a subscriber starts their access session. This enables users to receive notifications of problems with services or networks affecting them immediately, i.e. before they try to use the service or network respectively.

4. OVERVIEW OF THE TTRS COMPONENT

To provide any form of automated or semi-automated trouble management requires consideration of numerous issues. One of the key issues considered in developing the TTRS component was that it should, as far as possible, be independent of the services themselves. Of course, if the TTRS component is to help in performing automatic trouble resolution, i.e. without the need for human intervention, then this implies that the TTRS component and the service - or more typically the fault management part of the service - itself interwork to some extent.

Whilst it is true that the resolution of the particular troubles themselves is service specific, the nature of the interactions between the TTRS component and the different services is more generic. Specifically, the requirement is that the service supports some form of interface that allows for the exchange of trouble information. Put simply, the role of the TTRS component is to act as a central location where informations on troubles can be recorded, discovered and directed to the appropriate entities, e.g. fault management systems or other potentially remote trouble ticketing systems. The resolution of the problem itself - should it exist! - is up to the appropriate fault management system of the service. The architecture of the TTRS component is given in figure 4.

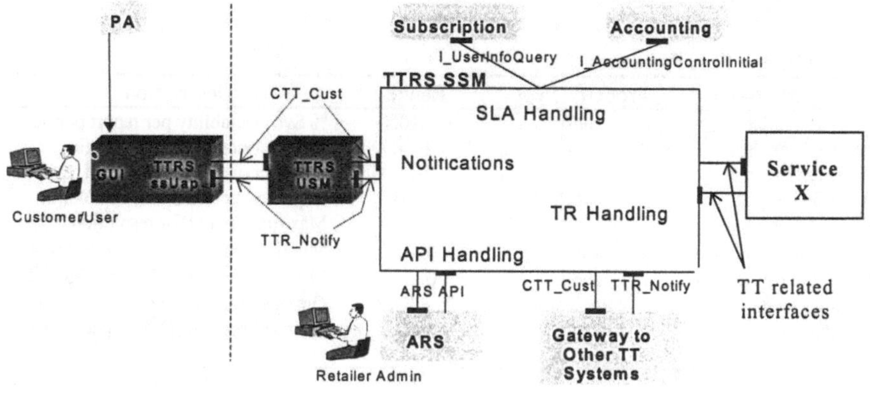

Figure 4. Architecture of the TTRS Component

The CCT_Cust and TTR_Notify interfaces depicted in Figure 4 are identical with those given in section 2. That is, the TTRS component implements the SMART IDL. More precisely, the SMART IDL has been implemented so as to provide a CORBA wrapper around a commercial trouble ticketing system: the Action Request System from Remedy [7]. This has been used predominantly as a database for trouble tickets, however it is quite possible to use it to perform other activities. For example, to send emails to users affected by troubles who do not currently have an active session. In addition the ARS was used to design the templates for trouble tickets held in the database. We consider the interface between the TTRS component and two example services in section 5.

The user interface to the TTRS service (the TTRS ssUAP) is itself represented as a traffic light with an additional window for writing trouble tickets or showing the information associated with existing troubles. The TTRS GUI traffic light is green when no troubles exist that affect this user; red when a trouble does exist that affects this user and yellow when troubles exist that need to be verified, i.e. the trouble has been resolved but that the user needs to verify that this is the case.

The TTRS component also interacts with subscription and accounting components. As stated previously, offering trouble management facilities in combination with service discounts for troubles incurred makes a service provider more competitive. To realise this the subscription component of the Y.TSP platform, has been extended to support service level agreements (SLAs). SLAs correspond to agreements made between customers and service providers upon the expected quality of the services subscribed to. When problems occur and are subsequently resolved (or possibly during the resolution process!), checks on the SLAs for the affected users are made, and if violated, can result in discounts being issued for the usage of that service. The most general SLAs common to all services are found in table 1.

Table 1. Generic SLA Properties

Property Name	Property Type	Range	Description
Service Availability	short	0...100	% Svc availability per report period
Activity time from	string	"0:00-24.00"	Daily service availability begin
Activity time to	string	"0:00-24.00"	Daily service availability end
Reporting period	short	28..31	Reporting period in days of month
Max Time to repair	float	0...∞	Max time (mins) for resolution
Time between failures	float	0...∞	Min time (mins) between failures
Minimum Errors	short	0.. ∞	Minimum number of errors allowed
Reliable Mode	Boolean	True/False	Optional transport mode supported
Maintenance time	float	0...∞	Maintenance time (mins) per report period

We note here though that violations of SLAs take two major forms: event based or non-event based. An example of an event based SLA is

TimetoRepair. Thus when a trouble occurs and subsequently resolved a check can be made immediately to see whether a violation has taken place. Service availability on the other hand checks for violations over some time period, e.g. a month. As such once a particular trouble has been resolved it will not necessarily lead to an immediate check for a violation.

We point out that certain of the SLA properties support various levels of violation. For example, when set by a subscriber the *TimeToRepair* property is given as a range of integer values, e.g. < 0, 1, 5, 10, 20, 60 >. Thus when a trouble occurs, the longer the trouble takes to be resolved the larger the violation that takes place. This in turn is used to influence the magnitude of the discount that is given. Thus here for example, if the service is fixed within 59 seconds then no violation takes place. If it is fixed within 61 seconds however, then a level-1 violation takes place etc. If it takes over 1 hour to repair then a level-5 violation takes place etc.

5. TROUBLE MANAGEMENT CASE STUDIES

5.1 The MusicShop Service

The MusicShop service has been developed and implemented as an e-commerce service integrated in the Y.TSP platform. It provides customers (musicians and consumers) with a multimedia store and retrieval system using Web-based technologies. It is implemented using the Oracle Application Server and the Oracle 8 RDBMS together with appropriate CORBA/Java technologies. The culmination of these technologies help provide a scaleable, reliable and manageable platform for hosting shared network applications such as data-driven multimedia content.

Musicians can upload music (or other media types, e.g. text, graphics etc) which can subsequently be downloaded by consumers. Musicians are charged for storing their music and general administration of their accounts, e.g. adding, deleting, renaming music folders etc, whilst consumers are charged for downloading music. The MusicShop service also supports a fault management interface which is used for transferring information related to trouble tickets associated with the MusicShop service. Specifically, the MusicShop supports a subset of the TTR_Notify interface described previously so that it can receive information on trouble tickets that have been created by TTRS due to potential faults with the MusicShop service. In addition, the fault management interface of the MusicShop service can be used to issue notifications about troubles it is currently having to the TTRS service via an interface offering a subset of the CTT_Cust interface.

Several scenarios have been developed to highlight the usage of trouble tickets in supporting service quality assurance using the MusicShop and

TTRS services. These scenarios illustrate both how trouble management systems can be applied for normal trouble resolution as well as highlighting how discounting for troubles incurred can be achieved. Examples of failures include the stopping of the http-listener so that the service effectively will not respond when the user attempts to interact with it or through the database being unavailable. We consider the former scenario.

To begin with we assume that the http-listener has been killed. Once the user has identified that the service is down they request that a trouble ticket is created using their TTRS-GUI. In this request they include the trouble object (MusicShop service), the trouble type (not responding) and other information that they might think is useful to the resolution process.

Upon reception of the request TTRS checks with the subscription component to check if this user subscribes to that service, and if so finds other affected users. The subscription information including any SLAs that are set for the affected users of MusicShop are returned. TTRS then creates trouble tickets for the affected users. Particularly important at this point is that a record of the time at which the trouble was initially identified is taken.

TTRS then issues a notification to the fault management part of MusicShop stating that it has created trouble tickets to a *not-responding* trouble with MusicShop. The fault management system then checks if such a problem exists, and issues an appropriate notification if it finds the trouble and begins working on resolving it. At this point TTRS sends notifications to all customers of the MusicShop service indicating the problem with the MusicShop service, thereby making their TTRS GUI go from green to red. The identity of the trouble tickets is also passed over at this time and can then be used for querying the state of the trouble by the user.

Once the fault management system of MusicShop has resolved the problem, i.e. the http-listener has been restarted, it sends an update to the TTRS component. Based upon the parameters of the notification, the TTRS informs customers that the problems is resolved and that they do/do not have to verify that it now works correctly, i.e. it now responds. At this point their TTRS GUI goes from red to yellow/green respectively. The time at which the problem was resolved by the fault management system is recorded. Once the users verify that the trouble no longer exists, the trouble tickets kept by TTRS for this particular problem are closed and the TTRS GUI shows green.

Once the trouble tickets are closed, TTRS checks whether any of the SLAs for the affected users have been violated. In this scenario, the *TimeToRepair* parameter is of particular relevance. The time difference between the trouble resolution and trouble identification times is taken. If this exceeds the value for an SLA limit as determined by the subscription information, then appropriate discounts are given, i.e. discounts are sent by TTRS to the accounting component indicating the reason for the discount.

5.2 The Dutch Auction Service

The Dutch Auction service implements a distributed online version of the real Dutch Auction which takes place daily in Aalsmeer in the Netherlands. Each day, the world biggest flower auction is held with flowers brought by farmers where they can be auctioned immediately.

The service itself supports two user roles, namely one *mediator* and an arbitrary number of (flower) *dealers*. The mediator (or auctioneer) controls the selling of the flowers. In contrast to an ordinary auction, where dealers bid with progressively increasing prices, a mediator's clock decrements the price until one of the dealers accepts the offer. At this point, all dealers clocks are stopped and reset in preparation for the next flower lot.

As a companion service to Dutch Auction, the Image Distribution Service (IDS) has been developed. This service allows dealers to get an image of the flowers currently being auctioned. Specifically, the mediator sends flower images to dealers currently using the Dutch Auction service.

The Dutch Auction and Image Distribution Service have stringent requirements on the transmission and delivery of the information they deliver, i.e. the bids and images respectively. These services need to ensure that the bids or images being transferred, are done so in a concurrent and timely manner, e.g. to ensure that all dealers have equal opportunity to bid for flower lots or to see images of the flowers. In this regard, the Dutch Auction service makes use of IP multicast and RSVP, the Internet Resource reSerVation Protocol. The result is the provision of a quality of service enhanced binding which allows a bandwidth guaranteed message transfer for, in this case, price bid information. The Image Distribution Service itself makes use of a Lightweight Reliable Multicast Protocol. It is important to note that this protocol can operate in both reliable or unreliable modes. Figure 5 illustrates the configuration of the image distribution service and the TTRS component.

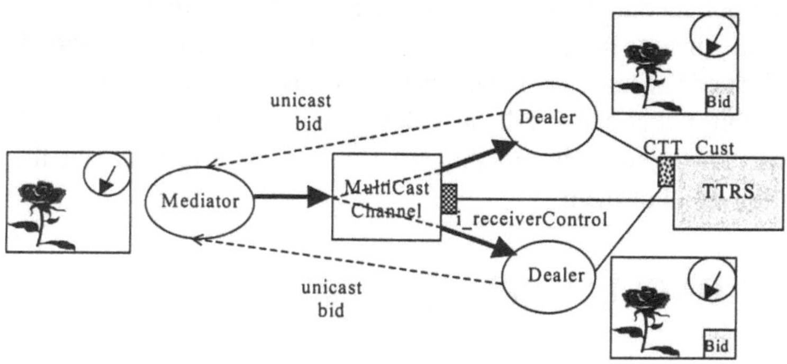

Figure 5. Configuration of Image Distribution Service and TTRS

Here the dealers receive images of flowers via the IP Multicast and make bids for them using a unicast connection. Normally the image distribution service runs in unreliable mode, however, should errors occur, e.g. due to too much network traffic with the connection between the mediator and the dealers, then it is possible to switch to (the more expensive) reliable mode. We note that errors are manifest as a lack of flower image.

In this scenario, requests to create trouble tickets for dealers suffering from connectivity problems, i.e. lack of flower image, are automatically sent to the CTT_Cust interface offered by TTRS. The parameters included in this are the trouble object (IDS), the trouble type (lack of image), and an interoperable reference (IOR) to the i_receiverControl interface. The TTRS component then interacts with the subscription component to see if a violation has occurred for this particular user. In this scenario the violations that are of interest are based upon the number of errors that have occurred and the time period over which they have occurred. Thus a dealer suffering from bursty types of failures, i.e. perpetual loss of images over a particular time period, can, provided they have selected the *true* option for the SLA switching to reliable mode parameter, switch transport modes. This is achieved through interacting at the i_receiverControl interface and changing the transport mode. This interface offers a single oneway operation that allows to set the different transport modes. Users who do not subscribe to this transport mode switching SLA parameter, are offered the option of upgrading their SLA values.

In comparison to the MusicShop-TTRS scenario various points are worth noting about the IDS-TTRS scenario. In this scenario the trouble tickets recorded in TTRS effectively have no useful history. That is, they do not require opening and clearing etc. Rather, here the trouble tickets act predominantly as timed counters which can be used to check the burstiness of recorded troubles. When a particular user has experienced more errors in a particular time period than that given in their SLA parameters, then the TTRS component automatically switches their mode for them.

Also in this scenario, the users troubles are isolated and largely independent of one another. Thus it is not the case that the request to create a single trouble ticket can result in TTRS creating many trouble tickets, e.g. for all users who subscribe to this service, and have to maintain the state information of all of them.

6. CONCLUSIONS

This paper has argued that to attract and keep customers, service providers require – amongst other things - better tools and techniques to increase their competitiveness. We have presented one example of such a trouble management tool together with case studies to illustrate its functionality. We have shown how this tool can be applied both to help in the trouble resolution process and as a basis for offering discounts to customers. The availability of such tools is increasingly necessary given the overlap between services and the networks they make use of, and the associated increase in complexity in the trouble management process.

The applicability of such tools to help minimise costs incurred is especially relevant to service providers. These cost minimisations can take several forms and offer the potential for numerous optimisations. For example, it is possible through such tools to attempt to avoid SLA violations from occurring, e.g. by prioritising the trouble resolution process for troubles that will result in a violation. Of course, a reduction in the trouble resolution process time is both beneficial to the service provider – so that more revenue can be generated, as well as to the customers themselves, since their services are available more often.

REFERENCES

[1] N. Agoulmine, D. Dragan, T. Gringel, J. Hall, E. Rosa, M. Tschichholz, Trouble Management for Multimedia Services in Multi-Provider Environments, Journal of Network and Systems Management, March 2000.

[2] Object Request Broker 2.0. Object Management Group, John Wiley & Sons, 1995.

[3] Covaci, S.; Dragan, D. Customer Care Solutions: Interoperable Trouble Ticketing Management Service, Proceedings of IS&N'97, Cernobbio (I), May 1997.

[4] Covaci, S.; Marchisio, L.; Milham, D. J., Trouble Ticketing X Interfaces for International Private Leased Data Circuits and International Freephone Service, Proceedings of 6th NOMS Conference, New Orleans, USA, February 1998.

[5] www.cs.ucl.ac.uk/research/flowthru/

[6] D.Lewis, A Development Framework for Open Management Systems; Interoperable Communication Network Journal, Baltzar Science Publishers, Q1, 1999.

[7] Remedy Corp - Part Number GSG-320-001, "Getting Started Guide"

[8] Reference Model of Open Distributed Processing, ITU-T Recommendations X.901..4 : ISO International Standard (DIS) ISO/IEC 10746-1..4, 1995.

[9] www.tinac.com

[10]H. Mulder (Ed.), TINA Business Model and Reference Points, Version 4.0, May, 1997

[11]A. Gavras, W. Takita, TINA DPE Architecture, Version 2.0b0, November, 1997

[12]F. Steegmans, TINA Network Resource Architecture, Version 3.0, February, 1997.

[13]P. Farley, R. Minetti, TINA-C Ret Reference Point Specifications, v1.0, Jan, 1998

[14]L. Kristiansen, TINA-C Service Architecture, Version 5.0, June, 1997

[15] TMF, Service Provider to Customer Performance Reporting Business Agreement, Issue 1.0, March 1997

[16]TMF, Customer-Service Provider Trouble Administration Information Agreement, Issue 1, March 1997

[17] Corba Interface Specification for Customer to Service Provider Trouble Administration, Revision 0.2, TM Forum, SMART Group, November, 1997

[18]Trouble Management function for ITU-T Applications – ITU-T Rec. X.790

[19]www.fokus.gmd.de/research/cc/platin/
[20]D. Dragan, T. Gringel, J. Hall, R. Sinnott, M. Tschichholz, W. Vortisch, Integrated Trouble Management to Support Service Quality Assurance in a Multi-Provider Context, Proceedings of 7th IS&N Conference, February 2000, Athens, Greece.

Chapter 15

Policies for Feature Interaction Resolution

Magdi Amer[1], Ahmed Karmouch[1], Tom Gray[2], Serge Mankovskii[2]
[1] Multimedia & Mobile Agent Research Laboratory, Dept. of Electrical and Computer Engineering, University of Ottawa, Ottawa, Ontario, Canada K1N 6N5, http://deneb.genie.uottawa.ca/
[2] Mitel Corporation, 350 Legget Drive, Kanata, Ontario, Canada K2K 2W7, http://www.mitel.com/

Key words: Policies, agent, fuzzy logic, conflict detection and conflict resolution.

Abstract: Telephone systems are marked by the provision of many hundreds of features. Conflict between these features is inherent as the actions of one feature can be in direct opposition to the aims of another feature. Most telecommunication service providers resolve the feature interaction problem by providing specific instructions in their management software. This approach suffers from the complexity of the resulting code and the difficulty of adding new features to the system. In this paper, we propose an agent-based architecture in which the actions of each agent are controlled by a set of policies. We also introduced the concept of fuzzy-policies, which are policies whose suitability for handling an event is calculated dynamically, based on the value of some fuzzy-variables. Conflicts are resolved using an arbitrator agent, which recalculates the suitability of the proposed actions of each agent and deduces the best action that satisfies the end user. The end user has the ability to add new policies, or modify the values of the fuzzy-parameters of the user-agent to alter the behavior of the system, thus obtaining a more personalized service.

1. INTRODUCTION

Telephone systems are marked by the provision of many hundreds of features that can be used to select the most appropriate way to dispose of a call for a user. Conflict is inherent in the provision of multiple features for a user. The actions of one feature can be in direct opposition to the aims of another feature. As a simple example the actions of the features Call Forward Busy and Call Waiting are usually given. With Call Forward Busy, a call to a user talking on the telephone will be directed to another telephone. In Call Waiting, the incoming call will be announce with a tone to the user who has an option of taking it directly while the incoming caller continues to hear audible ringing. These two features cannot be simultaneously active since they will direct incompatible outcomes for a call. With many hundreds of features, there is an explosion of such interactions, which must be detected and resolved.

The coming converged voice/data marketplace will create opportunities for many new features. As can be expected this will only make more difficult the problem of feature interaction since features can now depend on parameters from the data as well as the voice world. In addition, these new features will more than in the past depend on the current user context and so will become much more dynamic than in the past. Most current telecommunication service-providers resolve the feature interaction problem by providing specific instructions in their management software to handle scenarios where feature interaction may occur. This increases the complexity of the resulting code and the difficulty of adding new features to the system. Moreover, the output of the resolution between conflicting features is predefined.

This motivated several researches to be conducted using a multi-agents architecture for feature interaction resolution [Buh98 and Wei97]. These architectures proved to be more flexible and extensible than classical systems, but they still suffered from the fact that the output of the resolution between conflicting features is predefined. Thus, the end-user did not have a means of affecting the system decision to reflect the preferences of the end-user.

The essence of our work was to come up with an architecture that is flexible, extensible and capable of detecting and resolving conflicts. This was done by using an agent-architecture whereby the agent thinking and decision-making is controlled by policies. We also introduced the concept of the action suitability of a policy, which reflects how much an action proposed by a policy is suitable for handling an event. The action *suitability* is calculated dynamically based on '*fuzzy variables*', which reflects the user preference or the current state of the system. Fuzzy variables [Cox94 and Mot92] are variables that are neither 100% true nor 100% false. For example, a temperature of 80F may be considered as being rather hot, but it is also considered warm to some degree. To express this concept, we associate to the fuzzy variable *temperature* a value *hot* with a degree of truth (or degree of membership), since "*temperature is hot*" is partially true and partially false at the same time. Our algorithm for conflict resolution is based on the *suitability* of a policy.

Section 2 of this paper will give more details on the architecture of our system In Section 3; we will provide more details on the policy and their representation. Section 4 describes our graphical notation for scenario representation. Section 5 and 6 demonstrates the decision-making mechanism through some scenarios, and finally, our conclusion will be presented in Section 7.

2. SYSTEM ARCHITECTURE

Our work is based on the negotiation agent approach [Gri94], and extends the work described in [Buh98], in which a multi-agents-architecture was proposed to resolve feature-interaction conflicts dynamically. The major shortcoming of the previous system is that it associated fixed priority levels to different actions. When a conflict occurs, the action with the highest priority was chosen. The main goal of our system was to provide a more flexible mean for resolving conflicts.

In our system, we use MediaPath [Mit97b] to control the calls. MediaPath is a multimedia communications system based on open standards, manufactured by Mitel Corporation.

Agents communicate with each other using MicMac [Mit97a]. MicMac is a *tuple space* [Gel85], which is an instance of blackboard-architecture. In such architecture, entities communicate by publishing events on the blackboard. On the other hand, agents that are waiting for events place a query on the blackboard, in which the agents expresses the parameters and constraints of the event they are waiting for.

Each physical device is represented by a ***device-agent***. The ***device-agent*** contains the policies responsible for handling and controlling the requests and actions of this device. The ***device-agent*** may contain multiple ***feature-agents***. A ***feature-agent*** contains all the policies that control the feature to which the device is subscribed.

The end user may also be represented by a ***user-agent***, which contains all the policies representing the constraints that the system imposes to this user, depending on the role of the user. The ***user-agent*** contains also the policies and the parameters that express the preferences of the user. When a user is assigned a certain device, the device loads the policies that are relevant to this device from the ***user-agent***.

3. THE POLICY REPRESENTATION

As explained in Section 2, each device is associated with a ***device-agent***, which is responsible for controlling this device. Each feature to which the device is subscribed is represented by a ***feature agent***, that is considered as being a part of the device agent.

Each user is also represented by a ***user-agent***, which contains all the policies that the system imposes on the user, as well as those that express the user preferences. When a device is being employed by a specific user, the ***user agent*** is contacted, and all the policies that are relevant to the device are loaded into the ***device-agent***. The ***device-agent*** also contacts the ***Policy Server*** to load the policies that are related to this device.

There exist two categories of policies, obligation and authorization [Mar97].

Obligations are policies that are associated with triggering events. When one of these triggering events occurs, the agent containing the policies associated with this event will have the duty of realizing the actions described in these policies. A positive obligation policy (**Op**) describes the action that the agent wants to perform when the triggering event occurs, while a negative obligation policy (**On**) describes the action that the agent seeks to prevent. Authorization policies are not associated with an event. They represent the permission in case of Positive Authorization (**Ap**) and the refusal of permission in case of Negative Authorization (**An**).

Two levels of registration exist, ***notification*** and ***handling***. An agent may want to be notified of the occurrence of an action to change its internal state or to start a set of actions. An example of this is a cost agent, which is responsible for calculating the cost of using a service, that wants to be notified when a call is established, in order to start calculating the resulting cost. The second level of registration is ***handling***. In this case, the agent wants to have a monopoly on the control of an event. Notification policies are processed before handling policies, as they are usually used for setting the context of the event.

Policies can also be associated with fuzzy-variables that reflects how applicable the policy is to the current context of the application. This is called the '***suitability***' of the policy. A policy that does not have associated fuzzy variables is interpreted as being 100% suitable.

When an event occurs, the device-agent will test the conditions associated with the policies waiting for this event. The device-agent will execute all the actions suggested by notification policies whose conditions are true. The handling policies will be examined next. Actions of the handling policies whose conditions are true will be suggested to the agent. The device-agent will choose one of the actions suggested by the policies of the agent, or by other feature-agents contained in this device-agent. The choice of the agent will be based on the suitability associated with these actions. Note that *No-Action*, which means do-nothing, will always be suggested with a degree of suitability equal to zero. This is done to prevent actions with a negative degree of suitability from being carried out, which are the actions that received a denial of permission. More details on this mechanism will be provided in section 5 and 6.

```
<!ELEMENT Policy        (Event?, Subject+, Action, Target+, Condition*, Truth-Value?)>
<!ATTLIST  Policy        id           ID                    #REQUIRED
                         type         (O_p | O_n | A_p | A_n)  #REQUIRED >
<!ELEMENT Event         (Sender?, Receiver*, Parameter*)>
<!ATTLIST  Event         id           ID                    #REQUIRED
                         event-class CDATA                  #REQUIRED>
<!ELEMENT Subject       (Entity-Name | Object)>
<!ELEMENT Action        (Action-Name , Parameter*)>
<!ELEMENT Target        (Entity-Name | Object)>
<!ELEMENT Condition     (Variable-Name, Operator, Value)>
<!ELEMENT Truth-Value   (Parameter-Name*)>
<!ELEMENT Sender        (Entity-Name | Object | any)>
<!ELEMENT Receiver      (Entity-Name | Object | any)>
<!ELEMENT any           EMPTY>
<!ELEMENT Entity-Name   (#PCDATA)>
<!-- Entity-Name is used to reference other objects -->
<!ELEMENT Action-Name   (#PCDATA)>
<!ELEMENT Variable-Name (#PCDATA)>
<!ELEMENT Parameter-Name (#PCDATA)>
<!ELEMENT Object        (Parameter)*>
<!ATTLIST  Object        oid          ID                    #REQUIRED
                         name         CDATA                 #IMPLIED
                         symbol       CDATA                 #IMPLIED>
<!ELEMENT Operator      EMPTY>
<!ATTLIST  Operator type (lessThan | biggerThan | equals | notEqual |
                          biggerOrEqual | lessOrEqual) #REQUIRED >
<!ELEMENT Parameter     (Parameter-Name?, Value)>
<!ELEMENT Value         (String | Fuzzy-Value |  Boolean | Numeric)>
<!ELEMENT String        (#PCDATA)>
<!ELEMENT Fuzzy-Value   ((Label, Degree-of-membership)*)>
<!ELEMENT Label         (#PCDATA)>
<!ELEMENT Degree-of-membership (#PCDATA)>
<!ELEMENT Boolean       (true | false)>
<!ELEMENT Numeric       (#PCDATA)>
```

Figure 1. Policies' Document Type Declaration.

Policies are represented in our system using XML. XML was developed as a standard by the World Wide Web Consortium as a meta-language, which is a language for describing other languages. It is a subset of the

Standard Generalized Markup Language (SGML). Many scientific communities have decided to adapt XML to define a standard for information exchange, like the Mathematical Markup Language [W3Cc] and the Synchronized Multimedia Integration Language (SMIL) [W3Cd]. The XML representation of a policy is given in **Figure 1**.

Using the XML grammar, the operator "," is used to denote a sequence of elements. The operator "?" means that this element can occur at the most, once, or not occur at all. The operator "+" denotes one or more occurrence of an element, while "*" denotes zero or more occurrences of an element.

In our system, policies may contain one or more objects, called subject. A subject is the entity responsible for executing the action of a policy. A policy may also contain one or more targets, where targets are the entities on which the action is performed. A policy may have an event associated with it, as in the case of Obligation Policies. Constraints are the pre-conditions that must be realized for the policy to be applicable. Constraints can place conditions on the subject, target or mode of the policy. It can also place a condition on the state of the system.

To increase the readability of the paper, we will use a notation inspired from the work of [Mar97]. The general format of this notation is given below, with optional arguments written within brackets.

Policy_ID type [registration-mode] [trigger] subject(s) {action} target(s) [when constraint(s)] [truth-value]

Figure 2. The policy notation.

4. THE GRAPHICAL REPRESENTATION OF A SCENARIO

There exist many techniques for scenario representation, such as Use Case Maps [Buh95 and Aym98]. However, these methods remain at the abstract level and are incapable of showing a specific scenario, or they are deficient in explaining the decision paths of the scenario. For this reason, we needed to develop a graphical notation to represent both the state of the agent system and the decision paths for each step in the scenario.

Let's start by explaining the symbols used in this notation. The rounded rectangle is used to represent an **Agent**. We put the agent name in the top part of the rounded rectangle, while the attributes that are relevant to the scenario are shown in the bottom section, as shown in **Figure 3**.

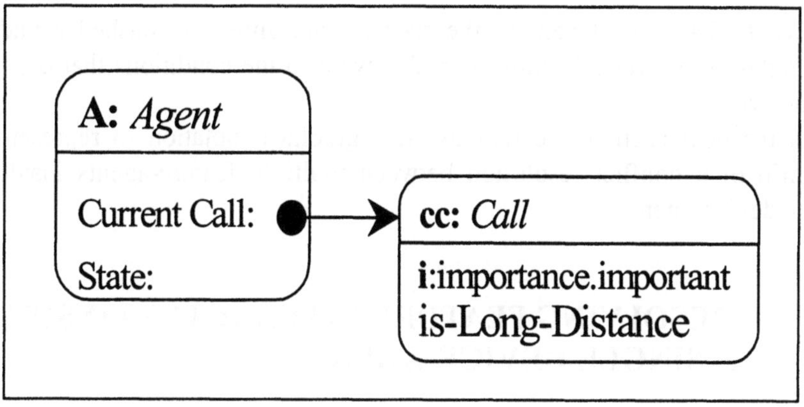

Figure 3. The Graphical Representation of an Agent.

A connector is used to represent an attribute that refers to another object. If no object exists, the connector will be grounded. Any agent, object or attribute can be represented by a symbol to facilitate referring to this element. For example, ***A.cc.ld*** refers to the parameter of Long Distance of the current call object of Agent **A**. To denote the specific label of a fuzzy variable, we append the variable name to the label name, separated by a dot. For example importance.urgent denotes the label "urgent" of the fuzzy variable "importance".

Actions are represented by ovals, within which we describe the action taken. An action may be associated with one or more groups of conditions. If any group of conditions is realized, the action will be proposed. A condition is symbolized by a hexagon and each group of conditions is connected to the action using the symbol of an encircled cross. The small circle bordering the conditions symbolizes that the action will be proposed if the condition is false. An action may have an associated truth-value that will be expressed by a down arrow callout, containing the parameters from which the truth-value is calculated. If no callout is shown, this means that the truth-value of this action is 100%.

The actions of notification policies are directly connected to the agent to symbolize that the agent will try to execute all the actions proposed. The actions of the handling policies are linked to the agent using an encircled dot to represent a decision point, as only one of these actions can be proposed. As explained previously, at each decision point we will have a ***No Action*** with an associated 0% truth-value.

If the Agent contains other sub-agents, this sub-agent will be connected to the Agent through an encircled dot to represent that the sub-agent can only suggest one action to the enclosing agent.

Events that are relevant to the scenario are shown in dashed rounded rectangles with arrows leading from the event to the conditions that depend on the event.

In the next section, we will use this graphical notation to represent a scenario of a conflict resolution between multiple feature-agents inside a single device-agent.

5. RESOLVING FEATURE INTERACTION INSIDE A SINGLE DEVICE-AGENT

In this example, we will explain how to use the agent architecture to resolve conflicts at the level of a single device agent. Let's suppose that we have a device agent containing three feature-agents, the Call Forwarding feature-agent (CF) and the Answer Call feature-agent (AC) and the Termination agent (T), which is responsible for accepting call requests. CW generates a call-waiting tone to alert the called party, whereas AC connects the calling party to an answering service. If **A** is already on the line when the second call comes in, should **A** receive a call-waiting tone or should the second call be directed to the answering service? The previous example is classified [Cam94] as a Single-User-Single-Component (SUSC), where the interactions occur because incompatible features are simultaneously in use by a single user in a single network component.

In this example, we assume that the user does not want to be interrupted during a long-distance telephone call. Thus, AC should take precedence over CW when the other call is long distance; otherwise, CW should take precedence over AC. The architecture of the agent is shown in **Figure 4**.

The device-agent contains the following notification policies.

- Pl_al **Op** [on *CR: CallRequest*] ***agent*** {set(0.6) } ***CR.importance*** [when not ***CR.isLongDistance***]
- Pl_a2 **Op** [on *CR: CallRequest*] ***agent*** {set(0.9) } ***CR.importance*** [when ***CR.isLongDistance***]
- Pl_a3 **Op** [on *CE: CallEstablishement*] ***agent*** {set(0.9* ***CE.importance***)} ***agent.currentCall.importance***

The policy *Pl_a1* says that when a local call-request is received, the agent should set the call importance (*C.importance*) to 60%. Rule *Pl_a2* says to set the importance of the call to 90% if the call is a long-distance call. Rule *Pl_a3* indicates that once the call has begun, the agent should reduce the importance of the call to 90% of its original value.

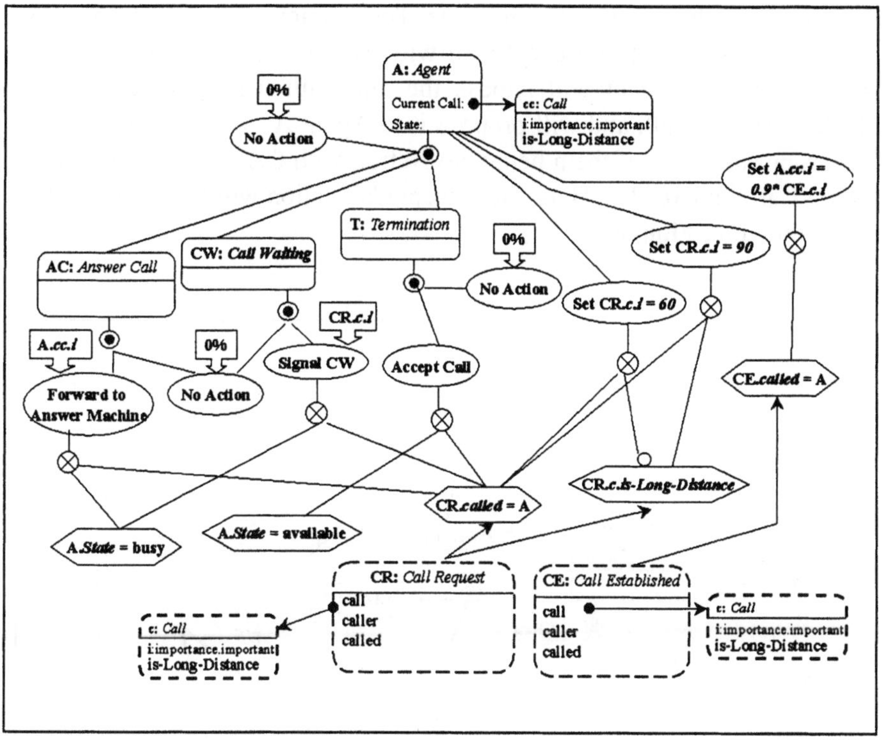

Figure 4. The Graphical Representation of the agent system.

The device agent contains three feature-agents, which are the Answer-Call agent, the Call-Waiting agent and the Termination agent.

The Answer-Call feature-agent contains the following policy.

- Pl_ac1 **Op** [on *CR: CallRequest*] ***agent.AC*** {forwardToAnswerMachine( ) } ***CR*** [when ***agent.state***= busy] [`Truth-Value` ***agent.currentCall.importance***]

This policy says that the call should be forwarded to the answer machine if the user is currently using the telephone. The truth-value of this rule will be equal to the fuzzy-parameter *importance* associated with the current call.

The Call-Waiting agent contains the following policy.

- Pl_cw1 **Op** [on *CR: CallRequest*] ***agent.CW*** {signalCallWaiting( ) } ***CR*** [when ***agent.state***= busy] [`Truth-Value` ***CR.importance***]

This policy says that the user should hear the call-waiting signal if the user is currently on the telephone. The suitability of this policy will be equal to the importance of the incoming call.

Let's suppose that a long-distance Call Request is sent to Agent A, while Agent A is available. From the Figure, we see that one action from the notification policies is true, which will set the importance of the Call Request to 90%. For the sub-agents Call Forward and Answer Call, only the

no action will be proposed, while the Termination Agent will have two actions proposed, that of Accept Call with 100% and No Action with 0%. The Termination Agent will choose the one with the higher truth-value, which is Accept Call, and will propose it to Agent A. Agent A will receive four proposals, three of them being No Action with 0% and the fourth being Accept Call with 100%. Thus, the Accept Call action will be executed. This step is shown in **Figure 5**.

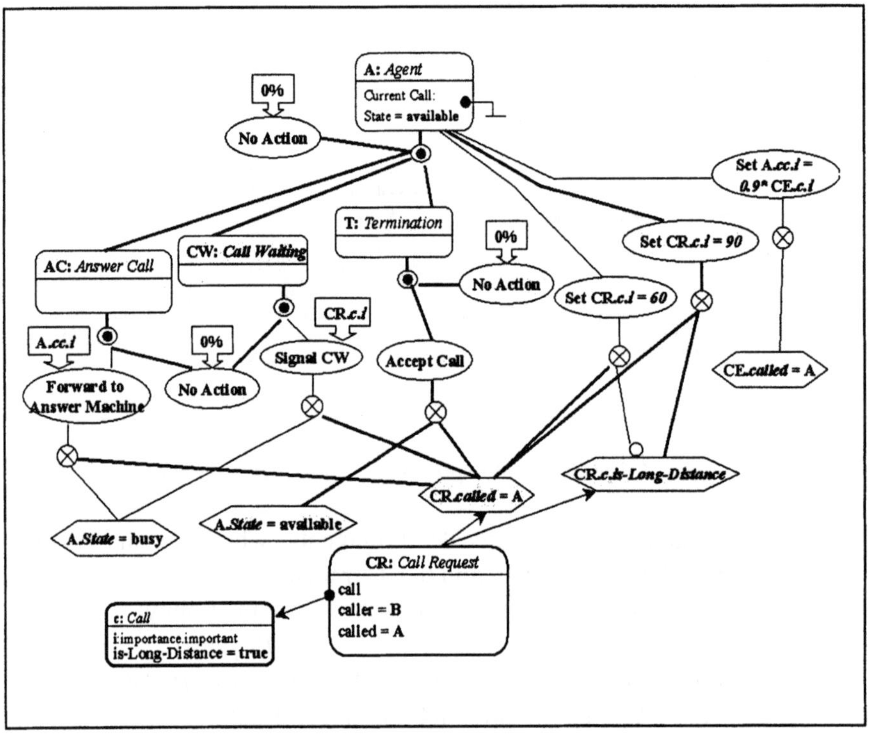

Figure 5. Step 1: Receiving a Long Distance Call Request.

Therefore, the call will be established and the Call Established event will be sent to Agent A, as shown in **Figure 6**. In this case, the state of Agent A will become busy, and the importance of the current call will be set to 90% of the importance of the call associated with the event, which is 90%. Thus, the importance of the current call will become 81%. On the other hand, all the sub-agents of Agent A will propose No Action with 0% truth-value, so no further action will be taken.

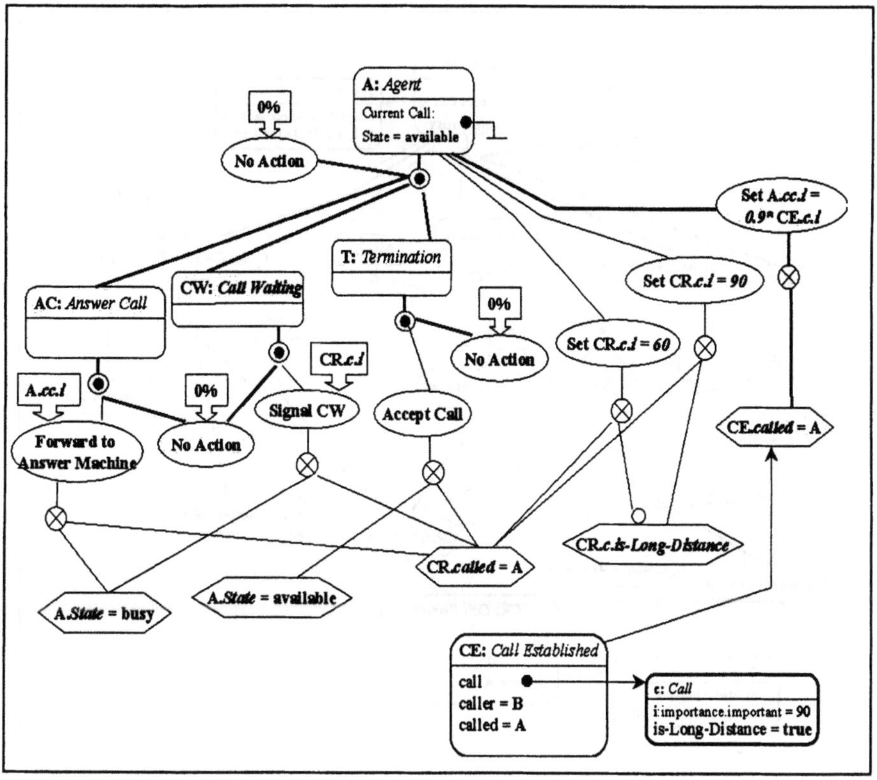

Figure 6. Step 2: Receiving a Call Establishment Event.

In **Figure 7**, we suppose that another long-distance call request was sent to the agent while the user was still busy. Similar to step one, only one notification policy will be triggered, which will set the importance of the Call Request to 90%. The Call Forward sub-agent will have to decide between No Action with a truth-value of 0% and Forward to Answer Machine action, whose truth-value is equal to the importance of the current call of agent A, which is 81%. Therefore, the Call Forwarding Agent will propose the latter action.

In the same way, the Answer Call will choose the Signal-Call-Waiting action, whose truth-value is equal to the importance of the current call, which is 90%. Agent A will have to choose between No Action with a truth value of 0%, Signal Call Waiting with a truth value of 90%, and Forward to Answering Machine with a truth value of 81%. So it will choose the action with the higher truth-value, this is the Signal Call Waiting.

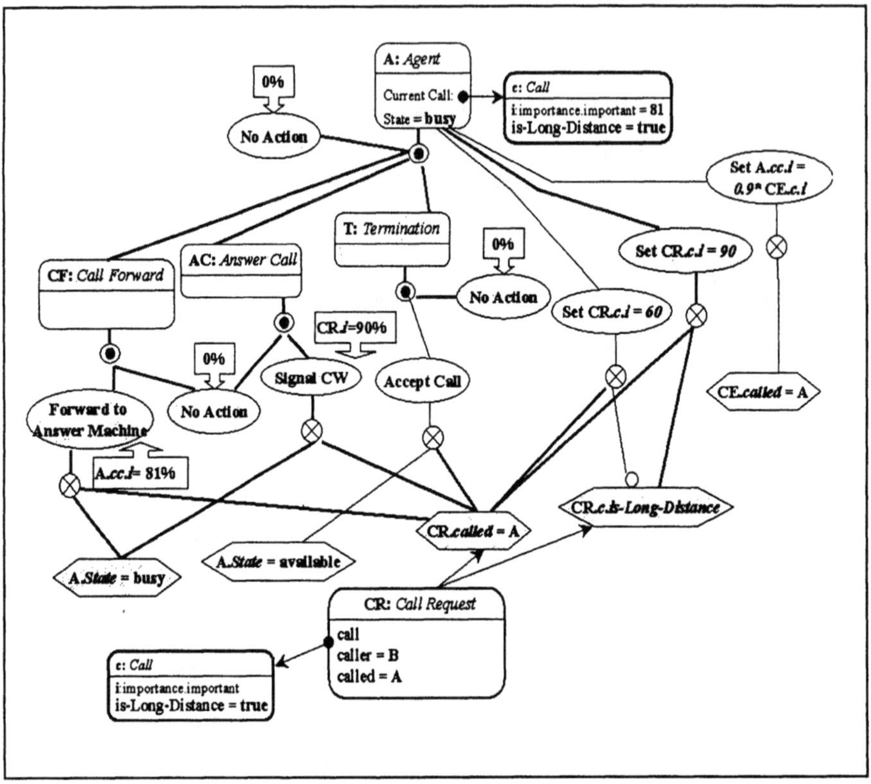

Figure 7. Step 3: Receiving a Second Long Distance Call Request.

In **Figure 8**, we suppose that the local telephone call was sent to Agent A while the user was still busy with a long distance telephone call. So only one notification policy will be triggered, which will set the importance of the Call Request to 60%. As in the previous case, the Call Forward Agent will propose the Forward to Answering Machine with a truth-value of 81%. The Answer Call Agent will propose a Signal Call Waiting with its truth-value equal to the importance of the Call Request, which in this case is 60%. In this case, Agent A will have to choose between No-Action with a truth value of 0%, Signal Call Waiting with a truth value of 60%, and Forward to Answering Machine with a truth value of 81%. The agent will choose the Forward to Answer Machine, as it is the action with the highest suitability.

As we've seen from this example, the graphical notations that we've developed help simplify the understanding of the system behavior, by allowing us to give a global view of each step of the scenario.

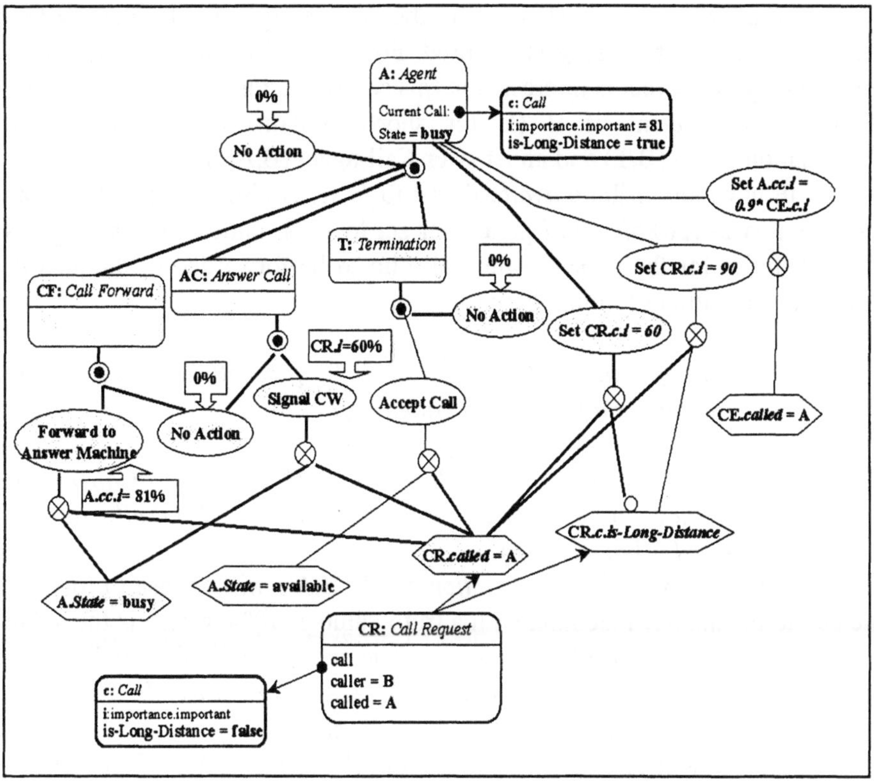

Figure 8 Step 4: Receiving a Local Call Request.

6. RESOLVING FEATURE INTERACTION BETWEEN MULTIPLE DEVICE-AGENTS

In this example, we will demonstrate how fuzzy logic can be used to resolve conflict between different device-agents.

Suppose that a user A wants to make a conference call with both B and C. A does not want any of the telephone calls to be forwarded to an answer machine or a secretary. Both B and C are not available. B has forwarded the telephone calls to the answer machine, while C has forwarded the telephone calls to a secretary. C has left a report with the secretary that should be read to the others in the conference call.

Let us take the first example of A sending a Call-Request (CR) to B.

The device-agent of A will post on the **blackboard** the following Call-Request.

- *CR.caller=A, called=B, id = X1*

The id of the call request is set by the originating agent to identify this request if further processing is required. For simplicity, let us suppose also that the only entity that is interested in this call request is the called agent B.

The device agent of B (B for short) contains two sub-agents, a Termination-agent (B.T) and an Answer-Call-agent (B.AC).

B will have an attribute *available* set to *false*. B also contains the fuzzy variable *forwardToVoiceMail* parameter set to 10%, indicating that it gives the possibility for the persons who call him to be forwarded to the answer machine, but without forcing them to.

B.T contains the following policies.

- *Pl_B_T1* ***Op*** *[on CR:CallRequest] B.T {acceptCall()} CR [when (B.available= true) AND (CR.called=b)]*

B.AC contains the following policy.

- *Pl_B_AC1* ***Op*** *[on CR:CallRequest] B.AC {forwardCall()} CR [when (B.available = false) AND (CR.called=B)] [Truth-value is (B.forwardToVoiceMail)]*

The output of B.T will be *no-Action*, while B.AC will choose to forward the call to the answer machine, with a truth value of 10%, which is the value of the *forward-To-Voice-Mail* parameter. The agent B will arbitrate between the output of the two modules and will choose the *forward-Call.* The result of this event will be a call forward on this call request.

The system will ask for the permission of the originator of this event, which is A.

A contains the following policy.

- *Pl_A1* ***An*** *A {forwardCall} CallRequest [Truth-value (A.noForward)]*

This rule indicates that A refuses to forward this call with a degree of truth equal to the value of the fuzzy input variable *no-Forward.* Let's suppose that A has set this parameter to 80%.

In this case, a No-Permission-Exception event will be generated, indicating that a suggested action has been refused. The **arbitrator agent** will be invoked to handle this event.

The **arbitrator agent** will recalculate the truth-value that was associated with the suggested action. The new value will be recalculated as equal to the original truth-value of the action multiplied by the weight that is accorded to the agent suggesting the action, minus the truth-value of the permission refusal action multiplied by the weight that is accorded to the agent refusing the action.

Suppose that we are using a passive arbitrator, which applies equal weights for every agent involved in the exception. In this case, the *forward-Call* truth-value will be evaluated to 10% - 80% = -70%.

B will re-evaluate the decision that it makes. Thus, B will have to choose between *No-Action* with 0% and *forward-Call* with -70%. B will choose the action with higher truth-value, so *No-Action* will be chosen. This means that the call will not be forwarded and the caller will receive a busy signal.

Now let us consider the case of A sending a Call-Request (CR) to C.

The device-agent of A will post on the **event-blackboard** the following Call-Request.

- *CR.caller = A, called = C, id = X2*

The device agent of C contains two sub-agents, a Termination-agent (C.T) and a Call-Forward-agent (C.CF).

C has the attribute *available* set to *false.* C contains also the fuzzy variable *forward-Call* set to 10%, to allow the calls to be forwarded to C's secretary.

C contains the following policy.

- *Pl_C1* ***Op*** *[on CR:CallRequest] C {set to 95%} forward-Call [when (CR.called=C) AND (CR.caller=A)]*

The previous policy is set by C to indicate that it is waiting for a call from A. When this call is received, the value of *forward-Call* is set to 95% to express the will of C to forward this particular call to C's secretary.

C.T and C.CF will contain the following policies.

- *Pl_C_T1* ***Op*** *[on CR:CallRequest] C.T {acceptCall} CR [when (C.available= true) and (CR.called=C)]*
- *Pl_C_CF1* ***Op*** *[on CR:CallRequest] C.CF {forwardCall( ) } CR [when (CR.called=C) AND (C.available = false)] [Truth-value (C.Forward-Call)]*

When the call request is received from A, C will set the value of *forward-Call* to 95%. The two sub-agents of C will suggest *no-Action* and *forward-Call* with a degree of truth of 95. Again, the system will ask for the permission of the originator of this event, which is A.

A will activate the rule *Pl_A1* and will refuse the call forward. A No-Permission-Exception event will be generated again and the **arbitrator agent** will be invoked again. The **arbitrator agent** will recalculate the truth-value associated with the suggested action. This value will be equal to 95% - 80% = 15%. Thus, the *forward-Call* action will still be the suggested action with the highest truth-value. The **arbitrator agent** will decide to override the objection of A and the call will be forwarded to C's secretary.

The examples shown in this section demonstrated the use of fuzzy policies. It explained how fuzzy logic constraints can be used to allow the user to alter the reaction of the system, depending on the user's own preferences, thereby giving the real control to the end user.

7. CONCLUSION

Telephony features may interact with each other in an undesirable or even an unacceptable way. Classical telephony systems resolve this problem by providing explicit instructions in their management software. With the increase in the number of features and the convergence of the worlds of data and voice, this approach is becoming less and less feasible.

In this paper, we proposed to use a multi-agents architecture to detect and resolve feature interactions. The end-user has the ability to alter the output of the interaction between the features by adding policies to the user-agent, thus making the system customizable.

ACKNOWLEDGEMENT

This work was partly funded by ***CITO***. It was the result of collaboration between ***the Multimedia and Mobile Agent Research Lab*** at the ***University of Ottawa***, the ***Network Computing*** group at ***National Research Council*** (NRC), and ***MITEL***.

REFERENCES

[1] Aho95 A. Aho, N. Griffeth, "Feature Interactions in the Global Information Infrastructure," in Foundations of Software Engineering, Washington, October 1995.

[2] Amy98 D. Amyot, N. Hart, L. Logrippo, "Formal Spe cification and Validation using a Scenario-Based Approach: The GPRS Group-Call Example". On-Line: http://www.sce.carleton.ca/ftp/pub/UseCaseMaps/wrroom98.pdf

[3] Bou94 L. G. Bouma, H. Velthuijsen, editors. "Feature Interactions in Telecommunications Systems", ISO Press, Amsterdam, 1994, 272 pp.

[4] Buh95 R. Buhr, " Use Case Maps: A New Model to Bridge the Gap Between Requirements and Detailed Design". On-Line: http://www.sce.carleton.ca/ftp/pub/UseCaseMaps/oopslaUCwrkshp.pdf

[5] Buh98 R. Buhr, D. Amyot, M. Elammari, D. Quesnel, T. Gray, S. Mankovski, "Feature-Interaction Visualisation and Resolution in an Agent Environment", in [Kim98], p.135-149.

[6] Cam94 J. Cameron, N. Griffeth, Y. Lin, M. Nilson, W. Schnure, H. Velthuijsen, "A Feature Interaction Benchmark for IN and Beyond". In [Bou94], p. 1-23.

[7] Che95 K. E. Cheng, T. Ohta, editors. "Feature Interactions in Telecommunications III", ISO Press, Amsterdam, 1995, 223 pp.

[8] Cox94 E. Cox, "The Fuzzy Logic Systems Handbook", AP Professional, Cambridge, 1994, p.624

[9] Din97 P. Dini, R. Boutaba, L. Logrippo, editors. "Feature Interactions in Telecommunications Networks IV", ISO Press, Amsterdam, 1997, 373 pp.

[10] Fra96 S. Franklin, A. Graesser, "Is it an Agent, or just a Program?: A Taxonomy for Autonomous Agents". In Proceedings of the Third International Workshop on Agent

Theories, Architectures and Languages. Springer-Verlag, 1996. On-Line: http://www.msci.memphis.edu/~franklin /AgentProg.html
[11] Gel85 D. Gelernter , "Generative Communication in Linda". In ACM Transaction on Programming Languages and Systems, Vol 7, No 1, p80-112.
[12] Gri94 N. Griffeth, H. Velthuijsen, "The Negotiating Agents Approach to Runtime Feature Interaction Resolution". In Feature Interactions in Telecommunications Systems, IOS Press, Amsterdam, May 1994.
[13]Kim97 K. Kimbler, "Addressing the Interaction Problem at the Enterprise Level". In [Din97], p13-22.
[14]Kim98 K. Kimbler and L.G. Bouma, " Feature Interactions in Telecommunications and Software Systems V", IOS Press, Amsterdam, 1998, 374 pp.
[15]Mar97 D. Marriott, "Policy Service for Distributed Systems", Master's thesis, Imperial College of Science Technology and Medicine, London, UK, 1997, p. 131
[16]Mit97a Mitel Corporation Micmac, 1997. Http://micmac.mitel.com
[17]Mit97b Mitel Corporation Mediapath, 1997. http://www.mitel.com/MediaPath
[18]Mot92 Motorola Corporation, "Fuzzy Logic Education program", The centre for Emerging Computer Technologies, Motorola Inc., 1992
[18]Nwa96 H. Nwana , "Software Agents: An Overview" In Knowledge Engineering Review, Vol. 11, No 3, pp.1-40, Sept 1996. On-Line: http://www.cs.umbc.edu/agents/ introduction/ao/
[19]Oht94 T. Ohta, Y. Harada, "Classification, Detection and Resolution of Service Interactions in Telecommunication Services", in L. G. Bouma, H. Velthuijsen, [Bou94], p. 60-72.
[20]W3Ca World Wide Web Consortium, " Frequently Asked Questions about the Extensible Markup Language". On-Line: http://www.ucc.ie/xml/
[21]W3Cb World Wide Web Consortium, " Extensible Markup Language (XML)". On-Line: http://www.w3.org/TR/PR-xml.html
[22]W3Cc World Wide Web Consortium, " Mathematical Markup Language 1.01 Specification". On-Line: http://www.w3.org/TR/REC-MathML/toc.html
[23]W3Cd World Wide Web Consortium, " Synchronized Multimedia Integration Language (SMIL) 1.0 Specification". On-Line: http://www.w3.org/TR/REC-smil/
[24]Wal97 N. Walsh. "A Technical Introduction to XML".On-Line: http://www.arbortext.com/Think_Tank/XML_Resources/ A_Technical_Introduction_to_XM/a_technical_introduction_to_xm.html
[25]Wei97 M. Weiss, T. Gray. "Experiences with a Service Environment for Distributed Multimedia Applications". In [Din97].
[26]Woo95 M. Wooldridge, N. Jennings, "Agent Theories, Architectures and languages: a Survey". In: ECAI-94 Workshop on Agent Theories, architectures and Languages. Eds.(Wooldridge M., Jennings N.). Springer-Verlag, Berlin, 1995. pp.1-39.

Chapter 16

A Piggybacking Policy for Reducing the Bandwidth Demand of Video Servers

Roberto de A. Façanha* (facanha@telemar-ce.com.br)
Nelson L. S. da Fonseca[&] (nfonseca@dcc.unicamp.br)
** Telecomunicações do Ceará, Telemar, Av. Borges de Melo, 1677, Fortaleza CE, Brazil*
& Institute of Computing, State University of Campinas, Campinas S.P., Brazil, PO Box 6176

Key words: Piggybacking, video-on-demand

Abstract: In video-on-demand systems, users expect to watch a film right after its selection. Nonetheless, such a short response time is feasible only if there is available bandwidth. In line with that, several techniques have been proposed to reduce the huge bandwidth demand on video servers. In this paper, we introduce the Piggybacking policy S^2, which adds a second level of optimization to the *Snapshot* policy. Moreover, we introduce a heuristic to reduce the complexity to generate the tree of superimposed video streams.

1. INTRODUCTION

Video on Demand (VoD) is one of the most promising applications in the future Broadband Integrated Services Digital Networks (B-ISDN). However, since it is extremely demanding on bandwidth, its availability in large scale requires the use of techniques for reducing this demand. Such techniques take into consideration the probability of a set of requests for popular videos (hot videos) getting to the system within a relatively short time interval, making it possible to provide a single video stream for all these requests.

Batching is a technique [1,2,3] which starts a new video stream by grouping all pending requests for a particular video in a certain time window. The main disadvantage of this technique is the delay introduced between the request and the beginning of exhibition of the video, which might lead the user to drop the request. *Piggybacking* is a technique [4,5,6] based on the fact that alterations of up to 5% in the display rate of a film are not perceptible to the user. In this way, requests can be promptly initiated and, by varying the display rate of subsequent video streams of the same film, we can superimpose the video streams as soon as they display the same frame. By doing so, we can provide only one video stream for all superimposed streams.

This paper presents the S^2 piggybacking policy. S^2 merges the video streams resulting from the use of the Snapshot policy on set of stream. Furthermore, a heuristic for reducing the complexity of generating merging trees is introduced. The Build Tree heuristic is $O(n^2)$ where n is the number of streams to be merged, whereas a solution via Dynamic programming is $O(n^3)$.

This paper presents the S^2 Piggybacking policy, which was discussed in the seminal paper [7]. The S^2 policy is a generalization of the *Snapshot* Algorithm policy [6]. Furthermore, we propose an heuristic for reducing the complexity of building merging tree.

The remaining of this paper is organized as follows: section 2 presents the *Snapshot* policy. Sections 3 and 4 introduce the S^2 policy and the proposed heuristic, respectively. Section 5 presents some conclusions. Appendix 1 presents the source code for the heuristic.

2. THE *SNAPSHOT* ALGORITHM

Snapshot is a Piggybacking policy [6] which attempts to minimize the number of frames shown for a given set of video streams. The computations performed by *Piggybacking* policies can be regarded as a binary tree (merging tree) in which the leaves correspond to streams, inner nodes are merges and the root is the final merge which forms the resulting stream of the set. Therefore, the number of possible trees generated by a set of streams constitutes the number of *Piggybacking* policies potentially optimal and is given by the $(n\text{-}1)$ st Catalan [6,7] number, that is:

$$Catalan(n-1) = \frac{1}{n!} \cdot \frac{(2n-2)!}{(n-1)!}$$

which implies that it is not viable to perform exhaustive search in an attempt to find an optimal strategy and its corresponding binary tree.

The *Snapshot* policy builds an optimal merging tree in the following way: consider a set of n streams of a single video, comprised of L frames, and their positions being given by $f_1, f_2, \ldots f_n$, where $f_1 \geq f_2 \geq \cdots \geq f_n$, at a given instant of time T. The minimum and maximum rates of frames per second of the video streams are denoted $S_{\min}$ and $S_{\max}$, respectively. Let i and j, where

$1 \le i < j \le n$, be two video streams. $P(i,j)$ denotes the merging position (in frames) at which streams i and j show the same frame. Since stream i has speed S_{min} and stream j has speed S_{max}, the merging position is given by:

$$P(i,j) = f_i + S_{min} \cdot \left(\frac{f_i - f_j}{S_{max} - S_{min}} \right)$$

Let $C(i,j)$ be the cost of a *Piggybacking* policy and $\mathrm{T}\ (i,j)$ the corresponding binary tree. The cost of a given stream is given by

$$C(i,i) = L - f_i$$

To minimize $C(i,j)$ it is necessary that the following optimality principle be satisfied: *there must exist a stream* k*, with* $i \le k < j$*, such that the left and right subtrees must also be optimal.* These subtrees are denoted by $\mathrm{T}\ (i,k)$ and $\mathrm{T}\ (k+1,j)$. The cost of the tree that corresponds to the set of streams $i,\ldots,j$ is given by:

$$C(i,j) = C(i,k) + C(k+1,j) + \max((L - P(i,j)),0) \quad (1)$$

Therefore, the optimal policy for the streams $i,\ldots,j$ contains subtrees $i,\ldots,k^*$ and $k^*+1,\ldots,j$, where

$$k^* = \arg\min_{i \le k < j} C(i,k) + C(k+1,j) - \max((L - P(i,j)),0)$$

In this way, the cost of the set of n streams, $C(1,n)$, can be computed in a bottom-up fashion by means of a dynamic programming algorithm.

The *Snapshot* policy merges video streams in an interval (*Snapshot* interval) given by $I = W/S_{max}$, where W is the optimal window size for the generalized simple merging policy [6]. At the end of the interval, the optimizing procedures described above are applied.

Sizing merging windows is an important issue. When the window size is large, a higher number of streams can be merged into a single stream, however, the mergings occur towards the end of the video. Conversely, when the window size is small, a reduced number of mergings are most likely to occur close to the beginning of the video. One can easily perceive that neither situation favors the reduction of the number of frames displayed by a set of streams. Aggarwal *et al.* [6] present an analytic method to optimize the size of the merging window to take into account the anticipated rate of new requests, assuming that these are modeled by a Poisson process.

3. THE S^2 PIGGYBACKING POLICY

The *Snapshot* policy was proposed to minimize the number of displayed frames from the streams contained in an interval I. The optimization involves only the streams that arrive during the last *Snapshot* interval. It is possible to verify that the smaller the interarrival time (higher rates) the smaller the optimal window size is, allowing for one or several optimal windows contained in the maximum catchup window. The possibility of merging two streams separated by at most W_m frames, where W_m is the maximum catchup window size, allows for additional gain on the top of those from the *Snapshot* policy. In line with that, we define a novel policy called S^2 which aims at optimizing the number of displayed frames of a set of streams in a modified maximal merging window, W'_m, $(W \leq W'_m \leq W_m)$ and not only in the interval I. The modified maximal window consists of an integer number of optimal windows, i.e., the modified maximal window has $\lfloor W_m / W \rfloor$ optimal windows. The S^2 policy introduces a second level of merges, in which streams resulting from the *Snapshot* intervals are merged.

The S^2 policy works as follows: it initially applies the *Snapshot* algorithm over streams which survive past the *Snapshot* intervals (as originally proposed). It subsequently applies once again the *Snapshot* algorithm on the streams resulting from the first step. We should emphasize that according to the definition of the *Snapshot* policy, the speed $S_{\max}$ is assigned to these resulting streams with the exception of the one generated by the first optimal window contained in W'_m. Another important aspect is that contrary to what happens in the first level of optimization, the point where the optimization procedure is subsequently applied does not have to occur at fixed intervals. These intervals are denoted by I_{S^2} and their duration is determined by the request pattern in each window W'_m (Figure *1*).

A natural generalization of the S^2 policy would be to consider n levels of optimization. Nonetheless, the gain that can be obtained by applying these additional levels are nearly null since streams in these levels are separated by roughly a number of frames which is close to W_m frames. Therefore, the introduction of extra merging levels would not produce effective reductions in the bandwidth demand.

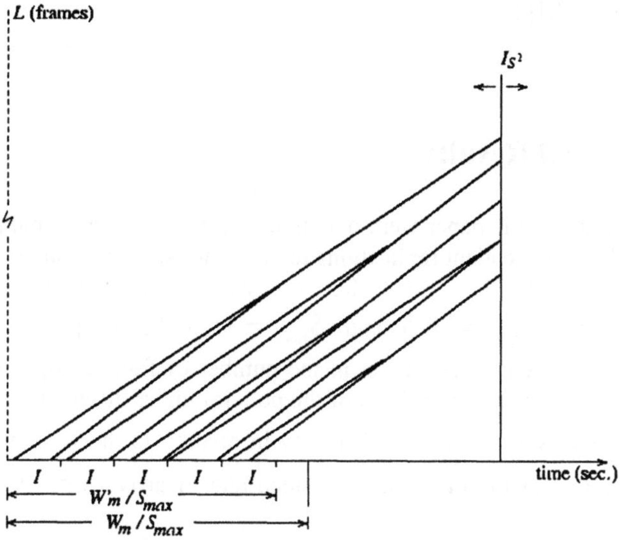

Figure 1. A scenario which shows that the last streams does not necessarily corresponds to the resulting streams of the last optimal window contained in the W'_m.

To evaluate the impact of introducing of a second level of optimization we compare S^2 to the original *Snapshot* policy and to the overall *Snapshot* policy which considers all the request for a given video without dividing time into *Snapshot* intervals. The objective function used in this paper differs from that considered in [6] which takes into account only the number of frames displayed after the interval I. Our objective function considers the total number of frames displayed for a set of streams, that is, it reflects the optimization that takes place in interval I as well as the ones which happen after that interval. So, the cost of exhibiting an individual stream is given by

$$C(i,i) = L$$

for a stream i, and is given by:

$$C(i,j) = Frames_I + (C(i,k) + C(k+1,j) - \max((L - P(i,j)),0))$$

for a set of streams $i,\ldots,j$, where the value of k is determined by Equation 1, and $Frames_I$ is the sum of frames displayed by the m discarded streams $Q(l)$, for $l = 1,\ldots,m$, before the end of the *Snapshot* intervals given by:

$$Frame_l = \sum_{l=1}^{m} Q(l)$$

3.1 Numerical Results

Results presented in this paper were obtaining via discrete-event simulation. We used the method of independent replications to compute confidence intervals with a confidence level of 95%. We assume that requests arrive according to a Poisson process and we use $S_{\min} = 28.5$ and $S_{\max} = 31.5$ frames per second. The graphics show the percentual reduction in the number of frames displayed. Three experiments were performed: i) analysis of the policies under different arrival rates; ii) analysis of the sensibility of the S^2 policy in relation to the window size; iii) analysis of the effect of the duration of a video and of arrival rate over the S^2 policy.

`Figure` *2* illustrates the behavior of the policies under different arrival rates, i.e., we vary the mean interarrival time from 15 to 500 seconds. We can see that as the mean interarrival time assumes higher values the benefit of adopting *Piggybacking* are reduced. For high rates, the introduction of a second level of optimization allows gains of up to 8% better than those achieved by the *Snapshot* policy.

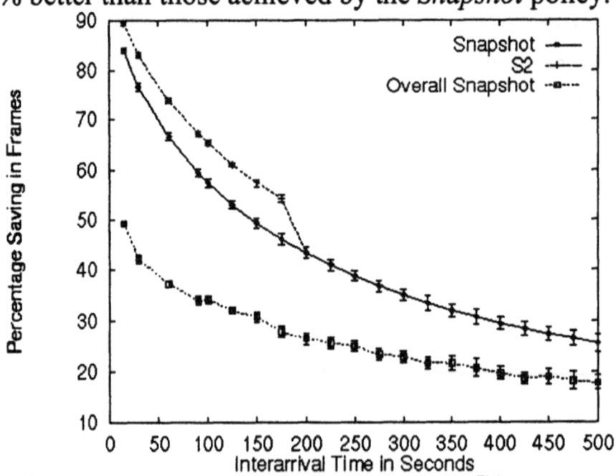

Figure 2. Comparison between the original Snapshot, the S^2 and the global Snapshot policies

`Figure` *3* shows the behavior of the S^2 policy as a function of the window size considering mean interarrival times of 30 seconds and a video duration of 2 hours. It can be seen that the S^2 policy is insensitive to the variation of the window size. This is due to the fact that the second level of optimization compensates eventual mergings not performed due to small catchup window. In other words, when the size of the catchup window is small, few merges occur at the first level of optimization which implies in a small reduction on the number of displayed frames. Since there is a second level of mergings, the resulting streams from the first merging level can still be merged leading to further reduction. As the catchup window gets close to maximum size, most of the gain comes from the first level of optimization.

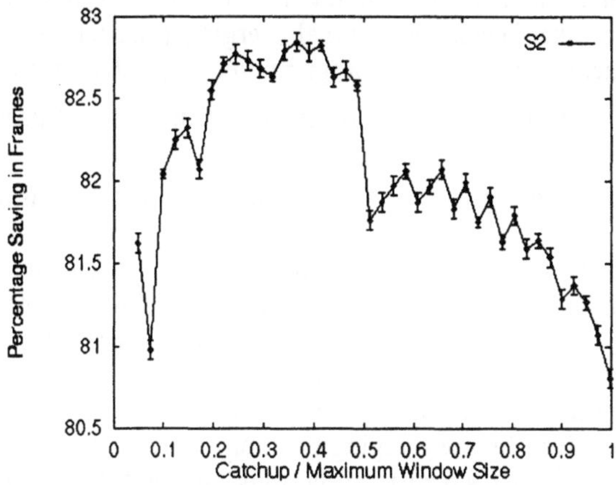

Figure 3. The impact of the window size on the percentage of nondisplayed frames under the S^2 policy

Figure 4 illustrates the impact of both the video duration and the mean interarrival time on S^2 savings. In this figure, we consider videos with duration from 30 minutes up to 4 hours and mean interarrival time from 15 to 500 seconds. It can be seen that we obtain greater reductions on the number of displayed frames for longer videos in conjunction with higher arrival rates. The reduction on the bandwidth demand varies from 9% for videos of 30 minutes duration and mean interarrival time of 500 seconds. The reduction can be up to 93% for long videos (4 hours of duration) and interval of arrivals of 15 seconds.

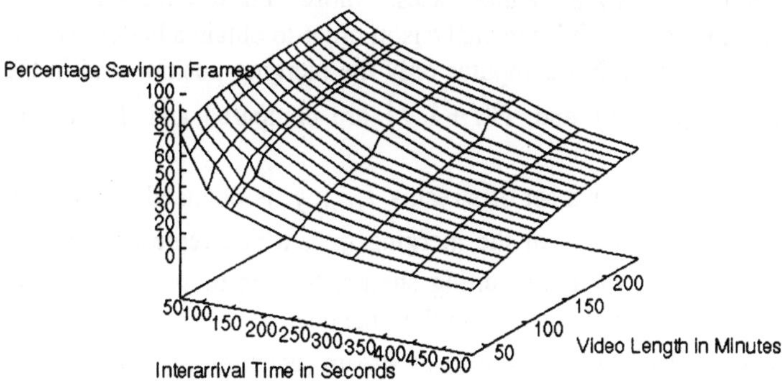

Figure 4. The impact of both request interarrival time and video length on the percentage of nondisplayed frames under the S^2 policy

4. REDUCING THE COMPLEXITY OF GENERATING MERGING TREE

If on one hand, the *Snapshot* policy is an attractive bandwidth reduction technique. On the other hand, the optimal merging (binary) tree of the optimal strategy is constructed by a dynamic programming algorithm whose complexity is $\Theta(n^3)$ [8], where n is the number of streams that survived past the *Snapshot* interval.

It was observed that merging trees can be built in a top-down fashion, differently from the dynamic programming algorithms which adopts a bottom-up fashion. In order to reduce it's complexity, we conceived a heuristic, called BuildTree, which is based on divide and conquer strategy. The basic principle of the heuristic consists in successively partitioning the set of streams to be merged in two groups and computing the costs of each subgroup until we obtain the merging tree. The division criterion used takes into consideration the segments of the merging tree (the number of frames of video given by the merging position) between the first stream and an intermediate stream (divisor stream) and between this and the last stream. In other words, we determine the stream that would represent the point of division of the set in potentially optimal subtrees. In this way, for a set of streams $i,\ldots,j$ there exists a stream k, with $i \le k < j$ which minimizes the absolute value of the difference the lengths of the segments $P(i,k)$ and $P(k,j)$, given by:

$$k^* = \arg\min i \le k < j\{|P(i,j) - P(k,j)|\}$$

After determining k^*, the original set of streams $i,\ldots,j$ is divided in the subsets $i,\ldots,k^*$ and $k^*,\ldots,j$, if $P(i,k^*) < P(k^*,j)$, or $i,\ldots,k^*-1$ and $k^*,\ldots,j$, otherwise, and then, the division criterion is re-applied to both. The algorithm considers two particular cases, whose entries have only two or three streams. These cases simplify the analysis in order to obtain a better performance. In the first case, the algorithm computes only the merging position and in the second, the segments $P(i,i+1)$ and $P(i+1,j)$ are compared and the larger value is discarded.

Therefore, the cost for a set of streams $i,\ldots,j$ is obtained through the sum of the lengths of the segments of the merging tree, which represents the cost of the $n-1$ streams except for the resulting stream. So, we must add the length of the resulting stream to this value to obtain the final cost.

In the process of determining the divisor stream (equation 5), $O(n)$ operations are sufficient. We can use the following property to reduce the number of operations carried out: "the divisor stream is the first whose merging segment from the initial stream to the one after the divisor is greater than the merging segment of the stream following the divisor up to the last stream." But, in the worst case, $k = j-1$, $O(n)$ operations still take place.

THEOREM: THE BUILDTREE HEURISTIC HAS COMPLEXITY $O(n^2)$.
PROOF: See [10].

For a better understanding of the heuristic consider the following example. Suppose that in a given *Snapshot* six streams are initiated and that at the end of it, their positions are: 3338, 3010, 2908, 2316, 1650 and 503. Let i be the first stream; k be the divisor stream and j be the last stream of the set so that the six streams are partitioned into the sets $(1 \leftrightarrow 4)$ and $(5 \leftrightarrow 6)$, $(i=1, k=4, j=6)$, given that the fourth stream is the first stream in which $P(i,k+1) \le P(k+1,j)$, resulting in the tree in `Figure` 5-a. In sequence we re-apply the algorithm over the subsets $(i=1,\ldots,j=4)$ and $(i=5 \text{ e } j=6)$. `Figure` 5-b illustrates the final merging tree built by the algorithm (in this case, the same as the tree generated by the dynamic programming algorithm).

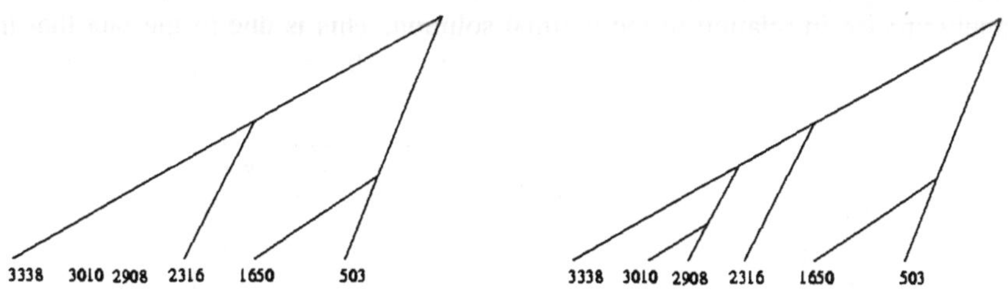

Figure 5. Construction of the merging tree by the heuristic BuildTree

In our experiments, the BuildTree heuristic was implemented using sequential search for the divisor stream in the set $i,\ldots,j$. The use of binary search would of course imply in considerable speed up. In Appendix 1, we show the heuristic as implemented in the C language.

4.1 Numerical Results

The proposed heuristic consists in an approximation of the optimal solution making it necessary to evaluate its precision in relation to the optimal one. With this intent, we execute both algorithms at the end of the *Snapshot* intervals to compare the costs of the generated merging trees. In this experiment, requests arrive according to a Poisson process and the mean interarrival time varies from 15 to 500 seconds.

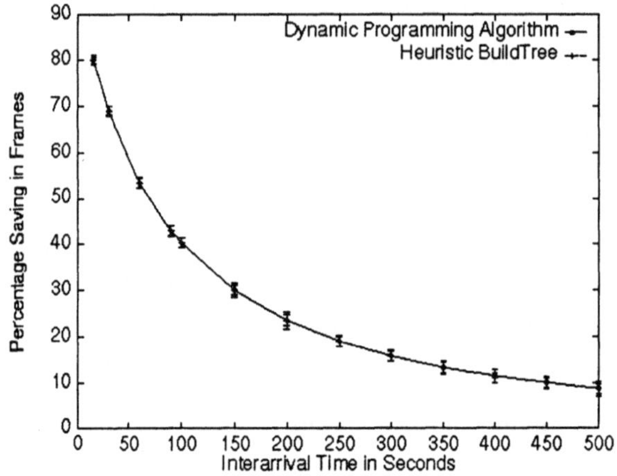

Figure 6. Comparison of the costs computed by the heuristic and by the dynamic programming algorithm

Figure 6 shows the mean percentual gain given by the dynamic programming algorithm and given by the heuristic. It can be seen that the BuildTree heuristic is extremely precise in relation to the optimal solution. This is due to the fact that in the majority of the cases the BuildTree heuristic obtains the optimal solution.

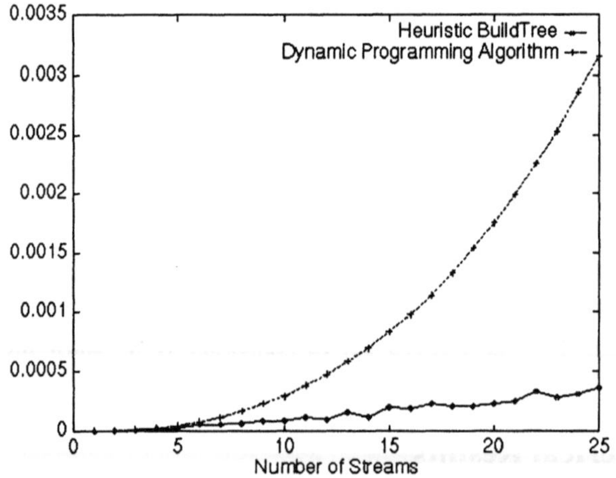

Figure 7. Comparison between the execution times of the heuristic and the dynamic programming algorithm

In *Figure 7*, instances from 1 to 25 streams (observed in other experiments) were fed as input for the computation of the execution times of both algorithms. We observed in all cases that the time in the execution of the algorithm was at most equal to the time of the dynamic programming algorithm.

5. CONCLUSION

This paper introduces the S^2 Piggybacking policy which adds a level of optimization in relation to the *Snapshot* policy. Moreover, we propose an approximate heuristic for reducing the complexity of building merging trees. We verified via simulations that the S^2 policy leads to less displayed frames than the *Snapshot* policy does. Furthermore, we note that the proposed heuristic is reasonably accurate, with a significantly lower complexity.

ACKNOWLEDGEMENTS

This work was partially supported by Pronex SAI, CNPq, and FAPESP.

REFERENCES

[1] Dan, D. Sitaram and P. Shahabuddin. "Dynamic Batching Policies for an on-demand Video Server". Multimedia Systems, Vol. 4, pp. 112–121, 1996.

[2] Dan, P. Shahabuddin, D. Sitaram and Don Towsley. "Channel Allocation under Batching and VCR Control in Video-on-Demand Systems". Journal of Parallel and Distributed Computing, Vol. 30, pp. 168–179, 1995.

[3] P. S. Yu, J. L. Wolf and H. Shachnai, "Design and Analysis of a Look-Ahead Scheduling Scheme to Support Pause-Resume for Video-on-Demand Applications". Multimedia Systems, Vol. 3, No. 4, pp. 137–149, 1996.

[4] L. Golubchik, John C. S. Lui and R. Muntz, Reducing I/O Demand in Video-on-Demand Storage Servers, In Proc. of ACM Sigmetrics, pp. 25–36, 1995.

[5] L. Golubchik, John C. S. Lui and R. Muntz, "Adaptive Piggybacking: A Novel Technique for Data Sharing in Video-on-Demand Storage Servers", Multimedia Systems, Vol. 4, No. 3, pp. 140–155, 1996.

[6] C. Aggarwal, J. Wolf and P. S. Yu, "On Optimal Piggyback Merging Policies for Video-on-Demand Systems", In Proc. of ACM SIGMETRICS, Vol. 24, No. 1, pp. 200–209, New York, May, 1996.

[7] Roberto de A. Façanha, N. L S. da Fonseca and P. J. de Rezende. "Reducing Bandwidth Demand on Video Servers", Reduzindo a Demanda de Banda Passante em Servidores de Vídeo (In Portuguese), In Proc. of the Second French-Brazilian Seminar on Distributed Systems: Multimedia Architectures for Telecommunications, Federal University of Ceará, Brazil and University of Versailles, France, pp. 308–319. Fortaleza, Ceará, 1997.

[8] M. Gardner, "Catalan Numbers", Scientific American, pp. 120–124, June, 1976.

[9] T. H. Cormen, C. E. Leiserson and R. L. Rivest, Introduction to Algorithms, The MIT Electrical Engineering and Computer Science Series, The MIT Press, 3rd edition, 1991.

[10] Roberto de A. Façanha, Nelson L. S. da Fonseca and Pedro J. de Rezende, The S^2 Piggybacking Policy, Multimedia Tools and Applications, Vol. 8, pp. 371–383, 1999.

APPENDIX

1. CODE OF THE BUILDTREE HEURISTIC

```
int BuildTree(int Posicao[], int i, int j) {
int n = j - i + 1, // Número de fluxos.
    k = i,         // Fluxo divisor.
    Custo = 0,     // Custo da árvore.
    MPik,          // Posição de Mesclagem dos fluxos i e k.
    MPkj;          // Posição de Mesclagem dos fluxos k+1 e j.
switch(n) {
    case 0:
    case 1:return(0);
    case 2:return(MergePosition(Posicao[i], Posicao[j]));
    case 3:MPik = MergePosition(Posicao[i], Posicao[i+1]);
           MPkj = MergePosition(Posicao[i + 1], Posicao[j]);
           return(((MPik > MPkj) ? MPkj : MPik) + MergePosition(Posicao[i],
                       Posicao[j]));
    default:
           while (MergePosition(Posicao[i], Posicao[k+1]) < MergePosition(Posicao[k+1],
                       Posicao[j]))
                k++;
           Custo = BuildTree(Posicao, i, k);
           Custo += BuildTree(Posicao, k + 1, j);
           Custo += MergePosition(Posicao[i], Posicao[j]);
           return(Custo);
      }
  }
```

Algorithm 1: Heurística de Construção da Árvore de Mesclagem.

PART SIX

Advanced techniques for Network Management

Chapter 17

INTEGRATING IP MULTICASTING IN STANDARD NETWORK MANAGEMENT PROTOCOLS

Ehab Al-Shaer Yongning Tang
Multimedia Networking Research Laboratory
School of Computer Science, Telecommunications and Information Systems
DePaul University
Chicago, IL 60604
(ehab,ytang)@cs.depaul.edu

Key words: Network Management, Group Communication, IP multicast, SNMP

Abstract: Supporting multi-point group communications in the network management platform is essential for improving scalability and responsiveness of management applications. With the deployment of IP multicasting as the standard infrastructure for multi-point group communications in the Internet, the integration of IP multicasting in SNMP becomes significantly important to achieve these goals. This paper presents a flexible, efficient and easy-to-integrate framework for integrating IP Multicast in standard SNMP agents. The proposed framework enables managers to re-configure the agents' group membership (e.g., join and leave) and the communication model (e.g., one-to-many, many-to-one and many-to-many) dynamically based on the application requirements. This framework exploits the advantages of IP multicasting without requiring any significant changes or performance overhead in the protocol or the agent architecture. Our ultimate goal is to promote the integration of IP multicasting as a standard service in SNMP agents.

1. INTRODUCTION

The Internet network management is based on client-server architecture where the manager is the client that sends the requests (e.g., *get* and *set*) and the agent is server that collects and reports information about *managed* objects. In network management environment, the manager may contact a group of agents frequently in local or remote domains to perform, for example, fault, performance or security management tasks. SNMP agents may also send same information to one or more managers for event correlation [Al-Shaer, 2000, Al-Shaer et al., 1999] or for increasing the information availability by redundancy. Therefore, the multi-point group communication between agents and managers is important to provide a scalable network management infrastructure that can handle large number of managed objects and agents. Compared to point-to-point (unicast) service, the multi-point group communication such as IP multicasting also reduces the latency and the number of messages of management tasks.

IP multicast is the de facto standard of multi-point group communication in the Internet. IP multicasting enables sending packets to a group of receivers without duplicating them at the source. Multicast packets are, instead, duplicated automatically by multicast routers that are close to the receivers in the network. Multicast receivers can start and stop receiving multicast packets dynamically through *join* and *leave* multicast operations respectively.

Current implementations of SNMP agents use only unicast for communication. Many approaches were proposed to support group communication in network management platform as a separate layer or a broker service between the manager and the unicast SNMP agents. However, in these approaches, the communication between the brokers and SNMP agents is still unicast and causes unnecessary overhead. In addition, they lack the dynamic reconfigurability of group communication at run-time. This paper presents a framework that facilitates integrating IP multicasting in standard SNMP agents with minimal changes in the agent implementation. The proposed framework is a manager-centric control which allows managers to reconfigure the agents' group membership and communication dynamically at run-time and based on the application demands.

This paper is organised as follows: Section 1 describes the design principles for efficient integration of multicasting in SNMP; Section 2 presents our proposed framework for integrating IP multicast in SNMP agents; Section 3 describes the framework implementation and benchmarking results; Section 4 discusses related work; Section 5 presents the summary and concluding remarks.

2. FRAMEWORK DESIGN PRINCIPLES

The main two issues that distinguish our approach are (1) using IP multicasting for multi-point communication, and (2) offering a flexible framework that can be customized by the network management application to construct any group management scheme. In this section, we describe the design principles to develop an efficient and flexible framework for integrating group communications in SNMP.

IP-multicast-based group communication. The employment of IP multicast in this framework is due to its availability in almost every network infrastructure today, and its scalability of large groups as no group information is required in the manager.

Dynamic Manager-centric Group Management. This framework enables SNMP agents and managers to join and leave multicast groups at anytime. It also enables network management applications (or managers) to configure the groups structure at run-time. This provides high flexibility while retains the simplicity of SNMP agents.

Flexible Group Communication. The framework also allows management applications (managers) to select the appropriate communication model (i.e., one-to-one, one-to-many, many-to-one and many-to-many) between SNMP managers and agents. In addition, managers may request the SNMP agents to reply to a unicast address, or a multicast address which may be used by multiple managers. This configuration can be performed dynamically at run-time.

Minimize overhead. It is important to show that this new feature cause no considerable performance overhead in SNMP agent. We discuss this issue in Section 3 and show that our framework has a negligible overhead in an SNMP agent.

Easy-to-integrate Framework. Our ultimate goal is to facilitate integrating IP multicasting in standard SNMP agents. To achieve this goal, the framework must require minimal code changes in SNMP agent implementations and other protocols.

3. A FRAMEWORK FOR INTEGRATING IP MULTICAST IN SNMP

In this section, we describe our approach and framework for extending the standard SNMP agent to support IP multicasting. The framework is designed based on the principles described in Section 1. In this framework, the integrated IP multicast service of an SNMP agent is completely configurable. The group information is stored in the MIB and used by the agents subsequently to establish the requested communication model and group organization. Therefore, the manager can request the SNMP agents to implement and dynamically change whatever group architecture suitable for the network management applications at run-time. In the following, we describe the extensions of the standard SNMP agent in order to support the IP multicast framework.

3.1 EXTENDING THE MIB WITH THE MULTICAST INFORMATION GROUP

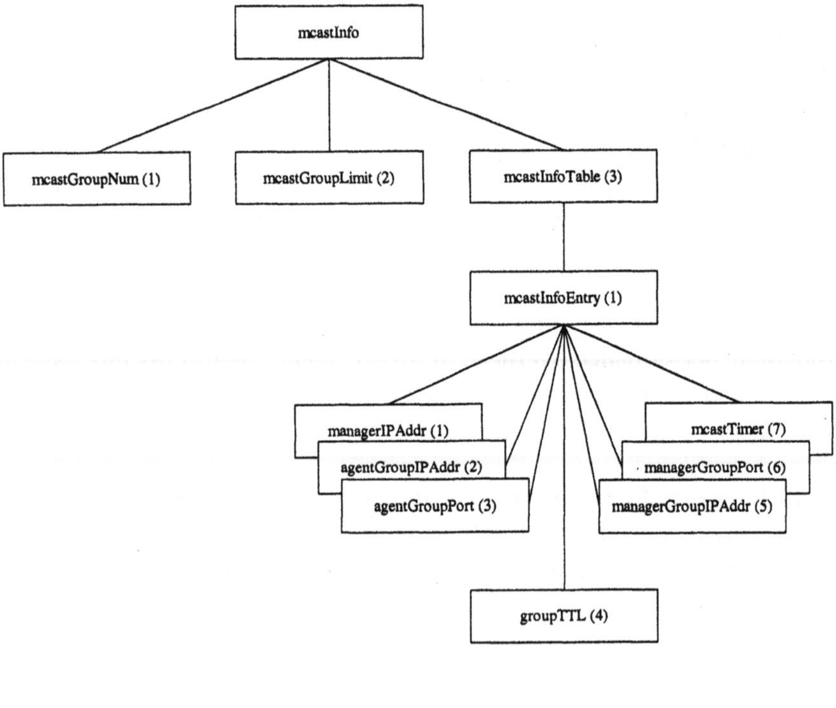

Figure 1. mcastInfo Group in Extended MIB II

A new MIB group, called Multicast Information Group (mcastInfo), was added to the standard MIB [McCloghrie and Rose, 1991] in order to store the multicast communication and group management information. Two classes of objects are supported in mcastInfo group:
(1) *Group Management Objects* that hold the address information. This class consists of **agentGroupIPAddr** and **agentGroupPort** to store the multicast IP address and port number, respectively, that the agent uses to join the multicast group. This class also contains **managerGroupIPAddr** and **managerGroupPort** that hold the IP address and port number of the manager multicast group respectively that the agent uses for the reply. If the **managerGroupIPAddr** or **managerGroupPort** is unspecified (i.e., NULL), then the agent replies to the unicast IP address of the requesting manager.
(2) *Group Communication Objects* that hold information about the communication parameters. It contains **groupTTL** object that specifies the TTL value in multicast packets, and **mcastTimer** object which indicates if the agent must use a randomized timer before sending the reply in order to avoid reply implosion in the manager [Floyd et al., 1997].

Because SNMP agent may interact with more than one manger that might request joining different multicast groups, the mcastInfo group is a table in which each entry contains the objects describe before and it is indexed by the IP address of the manager (**managerIPAddr**). Figure 1 shows the structure of the mcastInfo group table. It is important to notice that mcastInfo allows any manager to request the agent to join more than one multicast group.

Table 1. Example of Multicast Group Information in mcastInfo Table

MIP	AGIP	AGPort	MGIP	MGPort	TTL	Timer
128.39.2.2	224.4.4.4	5000	NULL	NULL	8	0
128.39.2.2	224.5.5.5	6000	224.33.2211	5555	8	0
128.39.5.5	224.8.8.8	7000	NULL	NULL	8	0

MIP: managerIPAddr AGIP: agentGroupIPAddr Timer: mcastTimer
MGIP: managerGroupIPAddr AGPort: agentGroupPort MGPort: managerGroupPort

Table 1 shows an example of mcastInfo table. In this table, the agent is requested by the manager of IP address 128.39.2.2 to join the following two multicast groups 224.4.4.4/5000 and 224.5.5.5/6000, and to send the replies to the manager IP address (i.e. using unicast because **managerGroupIPAddr** is NULL) for requests are received via 224.4.4.4, and to the multicast address 224.33.22.11 for requests received via 224.5.5.5. This tables also shows that the agent is requested by another manager of address 128.39.5.5 to join 224.8.8.8/7000 and to send the replies to the manager unicast address.

We propose inserting the mcastInfo objects in the SNMP Group of MIB II because this group extends the SNMP agent operations. However, for experimental purposes, we implemented mcastInfo group under *enterprise* of MIB II.

3.2 EXTENDING THE MASTER SNMP AGENT

The master SNMP agent (snmpd) must be extended in order to receive, process and respond to multicast set and get requests. We outline these extensions as follows:

(1) *Integrating the MIB Instrumentation Routines:* MIB II is extended to include mcastInfo group. This also involves implementing functions in the agent program that facilitate accessing the mcastInfo objects through *get* and *set* operations. This is a standard step in extending any MIB with new objects. As we describe later, the multicast operations (join and leave) are also invoked within these functions as a result of setting the mcastInfo objects.

(2) *Integrating Multicast Join and Leave Operations:* Multicast group

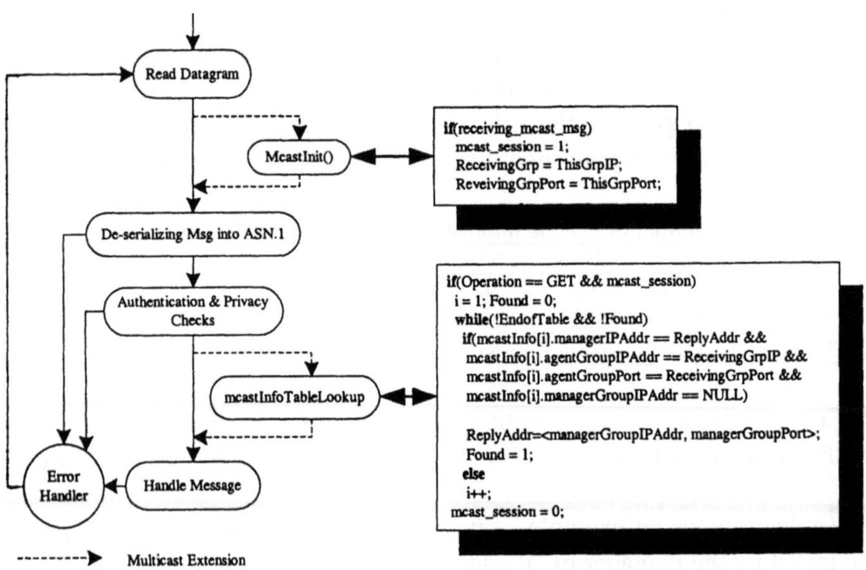

Figure 2. Extending the SNMP Agent Main Event Loop

membership is manipulated by agents through join and leave operations. The extended agent invokes these functions with the proper parameters obtained from the manager's *set* request. Another routine is also added to set the multicast TTL value in **groupTTL**. These functions (*joinGroup*() and *leave-*

Group()) are invoked by the agent as a result of setting the **agentGroupIPAddr** and **agentGroupPort** to valid values.

(3) *Extending the Agent Main Loop:* The main event loop of an SNMP agent basically consists of receiving, reading, authenticating and handling the incoming SNMP requests such as *get*, *set* requests. Figure 2 shows the main event loop of standard SNMP agents. The dashed lines in Figure 2 outline the multicast extension in the main event loop. As a result of reading the request, the IP address of the sending manager (*ReplyAddr*) is identified [Rose, 1994]. Because the proposed framework enables multicast managers to specify the type (unicast or multicast) and the multicast address of the response messages, the SNMP agent event loop was extended to search the mcastInfo table in order to find the reply address (*ReplyAddr*) before handling the message (Handle Messages in Figure 2). If the extended agent receives a multicast *get* or *get-next* request from a multicast group, then the address is stored in *ReceivingGrpIP* and *ReceivingPort* by *McastInit*() in Figure 2. The extended agent then uses *ReplyAddr*, *ReceivingGroupIP* and *ReceivingPort* to looks for a **managerIPAddr**, **agentGroupIPAddr** and **agentGroupPort** matching entry in the mcastInfo table.

The agent uses **managerGroupIPAddr** and **managerGroupPort** as the new end-point address for the reply message (*ReplyAddr*) if the corresponding entry is found. If **managerGroupIPAddr** or **managerGroupPort** is NULL (i.e., not specified by the multicast manager), then the unicast address of the sender is used for reply. This is performed by *mcastInfoTableLookup*() shown in Figure 2. It is important to notice that this function is invoked only if an SNMP request such as *get* or *get-next* is received via multicasting groups. This implies that this extension has no impact on the unicast-request path. In addition, later in Section 3, we measure and evaluate the overhead of these functions on SNMP agents.

(4) *Integrating Randomised Multicast Timer:* If the **mcastTimer** is *set* to a non-zero value by the manager, then the extended agent invokes the randomized timer function, *McastRandTimer*(), before sending the *get-reply* to the manger which puts the agent in waiting state for a very short random time. This disperses agents' replies over a short amount of time and avoid reply implosion at the manager host. We use the following exponentially weighted random timer as proposed [Handley, 1992] to estimate the random time value (*t*):

$$t = D_1 + RTT_{\max} * \log_2(2^{(D_2/RTT_{\max})} * X + 1)$$

such that X is chosen from the uniform random interval [0:1], $RTT_{\max}$ is the estimated maximum round trip time, and D_1 and D_2 are the ms time interval boundaries for *t*.

3.3 MANAGER-AGENT INTERACTION

Similarly, SNMP managers that intend to use the multicast service must also be extended to incorporate the mcastInfo MIB objects and may be the multicast join/leave operations. Since IP multicasting allow sending to a multicast group without being a member, managers are required to incorporate join and leave functions if they intend to receive multicast reply from agents.

This section shows the interaction of a multicast-capable manager and agent using this framework. They both were extended as described in Section 2.2. One of the key attributes of such framework is that it does not require using any proprietary protocol or request type to integrate group communication in standard network management protocols. Instead, our framework utilizes the existing SNMP *get* and *set* operations to configure and control group communications in network management environments.

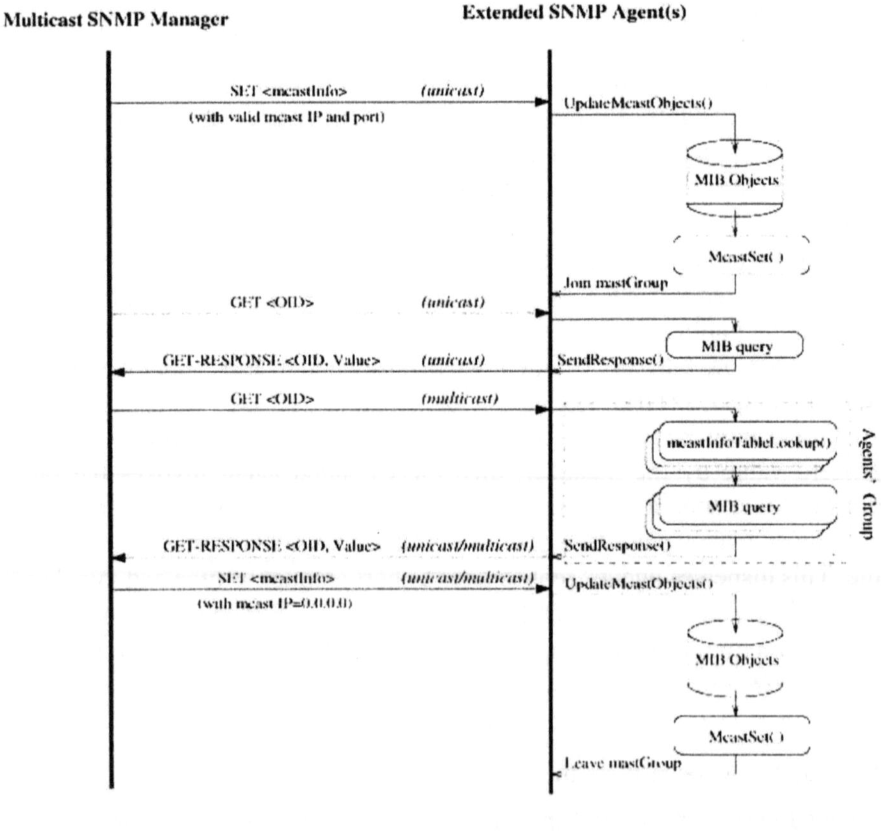

Figure 3. The Interaction of SNMP Manager and Agent Using the Multicasting Framework

Figure 3 shows the basic interaction between a manager and an agent using this framework. In this figure, the manager configures the group communication by setting the multicast information in mcastInfo objects of the agent. The manager sends this request as a unicast message to every agent to be involved in the group communication. Then, the agent reacts by updating the mcastInfo object in the MIB and joining the specified multicast group (*McastSet*() in Figure 3). On the other hand, the agent leaves the multicast group specified by **agentGroupIPAddr**, if this address is set to 0.0.0.0 by the manager. As a result of leaving a multicast group, the agents also deletes the corresponding entry in the mcastInfo table (using *McastSet*()). Since multiple managers can configure the same SNMP agent simultaneously, different managers may share the same mcastInfo objects of an agent. For this reason, the agent grants *set* requests only if they are received from the same managers that created the mcastInfo table entries. In other words, a manager can not change (*set*) mcastInfo objects that are created by another manager. If a manager attempts to set **agentGroupIPAddr** that is already in use (i.e., owned by another manager) or if the maximum number of groups indicated by **mcastGroupLimit** is exceeded, then the extended SNMP agent sends a *get-reply* to the manager reporting the problem.

If the manager uses the unicast session to communicate with an agent, then this agent always uses unicast to send the responses to the manager. Therefore, this multicast extension does not affect the standard unicast interaction between an SNMP agent and a manager which is an important design attribute of this framework. On the other hand, if the manager sends a multicast *get* or *get-next* request then, as shown in Figure 3, all agents joining this group fetch the requested values from the MIB and reply to the manager using the **managerGroupIPAddr** and **managerGroupPort** from the mcastInfo table as explained before. This means that manager can request agent to send responses as a unicast or a multicast message. Notice that even if a manager receives multicast replies from a group of agents, the manager can identify the agent IP address for each response. Therefore, the manager at any time can re-organize the agents multicast groups by requesting agents to leave a group (setting **agentGroupIPAddr** to 0.0.0.0) and join another group (setting **agentGroupIPAddr** to a valid multicast IP address).

4. THE FRAMEWORK IMPLEMENTATION AND BENCHMARKING

This framework was implemented and benchmarked in 100 Mbps switched Ethernet LAN environment that contains number of Sun Ultra workstations running Solaris 7 and 300Mz Pentium PCs running Windows

NT. We first extended the MIB II module to include mcastInfo group under *enterprises*. Figure 4 shows the extended MIB incorporated in the MG-Soft MIB browser (http://www.mg-soft.si/download.html). We use the implementation of SNMP agent from University of California at Davis (http://www.ucd-snmp.ucdavis.edu) as case study for developing this framework. The extended (multicast) SNMP manager is implemented using SNMP++ [Mellquist, 1997]. The implementation details of both the agent and the manager, and multicast integration plan is described in [Al-Shaer and Tang, 2000].

Three simple steps are required to integrate IP multicast service into a standard SNMP agent: (1) extending the MIB module to include mcastInfo group, (2) inserting *McastInit*() and *mcastInfoTableLookup*() functions in the proper places as specified in Figure 2, and (3) compile and link the source code of the SNMP agent with the multicast files and the extension library

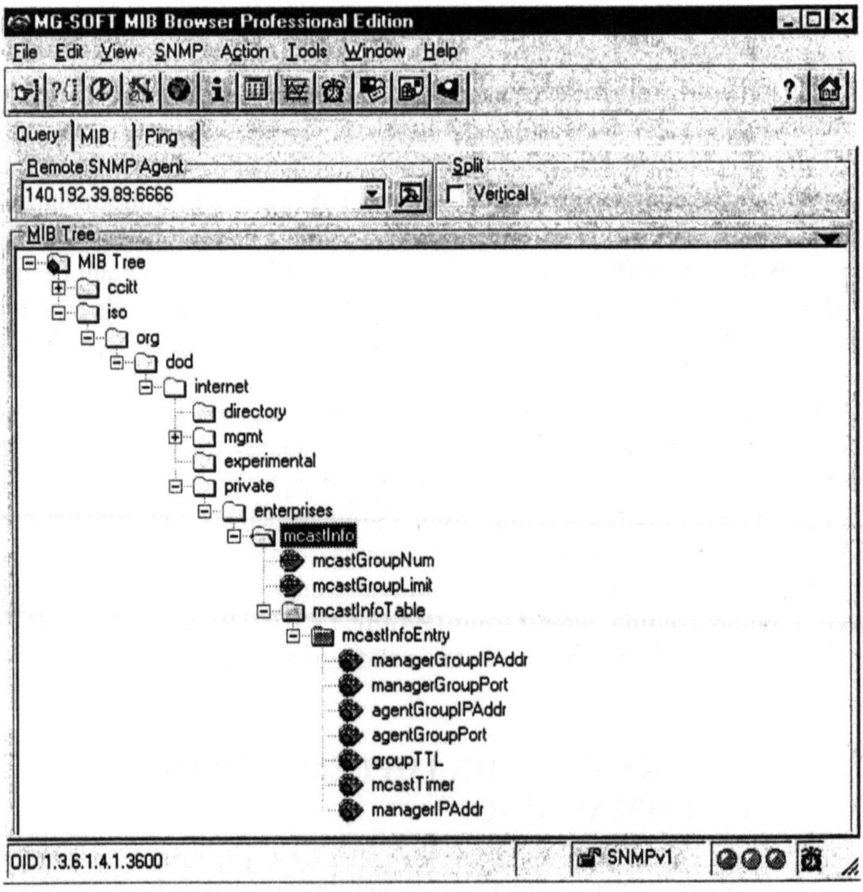

Figure 4. A MIB Browser Incorporating mcastInfo Group

(libmcastsnmp.a). The source code is available from http://www.mnlab.cs.depaul.edu/mcastsnmp/.

We conducted number of experiment to measure the overhead of this framework on SNMP agents. Figure 5 shows the response time of both standard SNMP agent using unicast and extended SNMP using multicasting agent after sending 80 different SNMP *get* requests (different OIDs) for 15 times each. Figure 5 shows that the difference in the response time between the standard and the extended agents is negligible (0 to 4 ms per response).

5. RELATED WORK

Our related-work study focuses on the systems that attempt to develop or use group communication in network management environment. A framework for using IP multicast group communication with SNMP is proposed in [Schonwalder, 1996]. This framework provides a primitive group membership structure and an election algorithm to choose a master agent. The SNMP trap messages over IP multicast are also utilized to exchange control

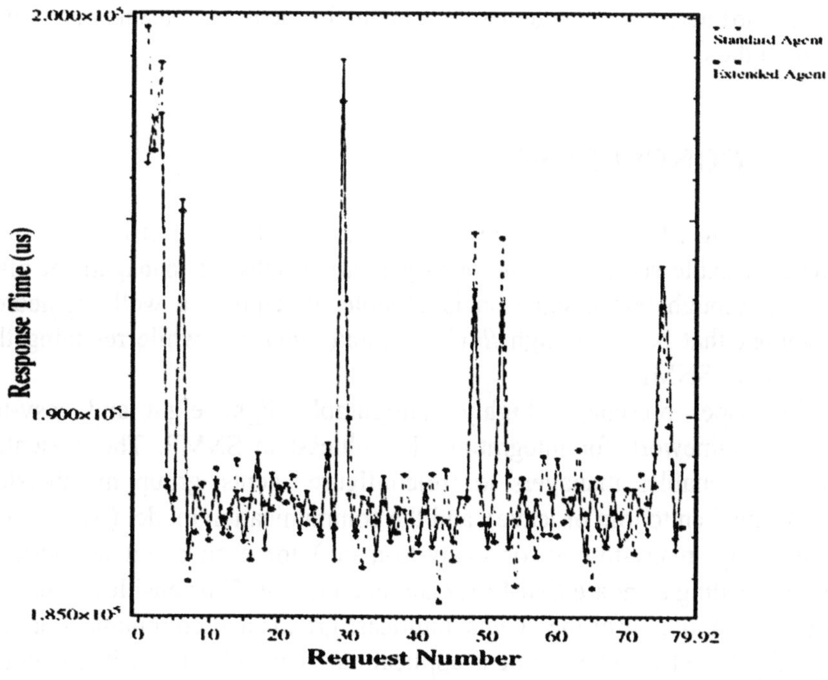

Figure 5. Response Time of the Standard and Extended SNMP Agents

information. Although this framework supports autonomous SNMP agents that use IP multicasting, it does not support an efficient framework for integrating IP multicasting in SNMP because it lacks the flexibility of multicast group and communication model re-configuration, and it requires a significant changes in the SNMP agent implementation.

In [Lee et al., 1995], a reliable group communication protocol for distributed management systems is proposed using a hierarchy of servers. However, it seems that this architecture developed over unicast connections, instead of employing IP multicasting, to emulate group communication. Furthermore, this work does not address integrating multicasting in SNMP agents.

Another group communication infrastructure based on Transis [Amir et al., 1992] group communication system was described in [Amir et al., 1996]. It is used for software installation, simultaneous remote execution and configuration in distributed systems management. A similar framework is proposed in [Parnes and Schefstorm, 1999] that uses IP multicasting for control and management of distributed applications. Although these systems show the advantage of using group communication in distributed management, they do not support an efficient framework for integrating IP multicasting in the standard network management protocols for the same reasons mentioned before in the first related work.

6. CONCLUSION

Integrating IP multicasting in SNMP agents is the most efficient way to provide scalable services in the next-generation network management platforms. Although this integration is feasible, it requires a well-engineered framework that provides high *flexibility*, and *efficiency* while retaining the *simplicity* of SNMP.

This paper presents a highly configurable, lightweight and easy-to-integrate framework for integrating IP multicast in SNMP. The presented framework enables managers to (re)configure agents' group membership (i.e., what/when to join or leave), and the communication model (one-to-one, one-to-many, many-to-one, or many-to-many) for each agent at run-time through sending standard SNMP *set* and *get* requests. This enables managers to organize and control the agents' multicast groups and communication dynamically based on management applications demands, and without introducing a new proprietary protocol. The benchmarking results show that our extended SNMP agent incurs a negligible performance overhead when compared with the standard one. The framework is implemented using UCD

SNMP agent and SNMP++, as an example, and available at http://www.mnlab.cs.depaul.edu/mcastsnmp/.

Our future research agenda includes enhancing the table mcastInfo lookup mechanism, improving the randomized timer technique to consider the group size and the location based on the network addresses.

REFERENCE

[Al-Shaer, 2000] Al-Shaer, E. (2000). Active Management Framework for Distributed Multimedia Systems. *Journal of Network and Systems Management* (JNSM), 8(1).

[Al-Shaer et al., 1999] Al-Shaer, E., Abdel-Wahab, H., and Maly, K. (1999). HiFi: A New Monitoring Architecture for Distributed System Management. In Proceedings of International Conference *on Distributed Computing Systems* (ICDCS'99), pages 171--178, Austin, TX.

[Al-Shaer and Tang, 2000] Al-Shaer, E. and Tang, Y. (2000). Implementation of SNMP Agent Multicast Extension, TR-2000-1. Technical report, School of Computer Science, Telecommunications and Information Systems, DePaul University.

[Amir et al., 1996] Amir, E., Breitgand, D., Chockler, G., and Dolev, D. (1996). Group communication as an infrastructure for distributed system management. In Proceedings of *Third International Workshop on Services in Distributed and Networked Environments*, pages 84--91.

[Amir et al., 1992] Amir, Y., Dolev, D., Kramer, S., and Malki, D. (1992). Transis: a communication sub-system for high availability. In IEEE Workshop on *Fault-Tolerant Parallel and Distributed Systems*, pages 76--84.

[Floyd et al., 1997] Floyd, S., Jacobson, V., Liu, C., McCanne, S., and Zhang, L. (1997). A Reliable Multicast Framework for Light-weight Sessions and Application Level Framing. IEEE/*ACM Transactions on Networking*, 5(6):784--803.

[Handley, 1992] Handley, M. (1992). Multicast address allocation protocol (AAP).

[Lee et al., 1995] Lee, K.-H., Lee, J.-K., and Kim, H.-S. (1995). A multicast protocol for network management system. In Proceedings of *IEEE International Conference on Information Engineering*, pages 364 --368.

[McCloghrie and Rose, 1991] McCloghrie, K. and Rose, M. (1991). *Management Information Base for Network Management of TCP/IP-based internets*: MIB-II, RFC1213.

[Mellquist, 1997] Mellquist, P. E., editor (1997). SNMP++: *An Object-Oriented Approach to Developing Network Management Applications*. Prentice Hall PTR, Roseville, CA, (http://tachyon.rose.hp.com/snmp++/).

[Parnes and Schefstorm, 1999] Parnes, P. and Schefstorm, D. (1999). Real-Time Control and Management of Distributed Application using IP-Multicast. In Proceedings of the IEEE/IFIP *International Symposium on Integrated Network Management (IM),* pages 901--914.

[Rose, 1994] Rose, M. T. (1994). The Simple Book*: An Introduction to Internet Management*. Prentice Hall, Englewood Cliffs, NJ.

[Schonwalder, 1996] Schonwalder, J. (1996). Using multicast-SNMP to coordinate distributed management agents. In *Proceedings of Second IEEE International Workshop on Systems Management*, pages 136 --141.

Chapter 18

Reducing the Response Time in Network Management by Using Multiple Mobile Agents

Marcelo G. Rubinstein[†,✤,1], Otto Carlos M. B. Duarte[†], and Guy Pujolle[✤]
† Grupo de Teleinformática e Automação, Universidade Federal do Rio de Janeiro, COPPE/EE, CP 68504, 21945-970, Rio de Janeiro, RJ, Brazil; ✤ Laboratoire PRiSM, Université de Versailles, 45, Avenue des Etats-Unis, 78035, Versailles, Cedex, France

Key words: Mobile Agents, network management, and performance evaluation

Abstract: This paper aims at investigating mobile agent tradeoffs in network management by performing management task simulations. Results show that the mobile agent performs better than the SNMP when the number of managed network elements ranges between two limits: an inferior bound, associated with the number of messages that pass through backbone links, and a superior bound, related to incremental size of mobile agent. Furthermore, if the network is divided into several domains and one mobile agent is assigned to each domain, instead of using a single mobile agent to manage the whole network, the results show that the response time decreases up to 90% when using 8 mobile agents.

1. INTRODUCTION

Most network management systems use SNMP (Simple Network Management Protocol) [1] and CMIP (Common Management Information Protocol) [2] protocols, which are based on a centralized paradigm. These protocols use the client-server model, on which the management station acts as a client that provides a user interface to the network manager and interacts with agents, which are servers that manage remote access to local information stored in a Management Information Base (MIB).

[1] Grant holder from CNPq - Brazil

The operations available to the management station for obtaining access to the MIB are very low level. This fine grained client-server interaction, called micro-management, generates an intense traffic that overloads the management station [3], resulting in scalability problems. But at the same time that the networks grow, the computational power of the managed element increases, making possible to perform management functions in a distributed way [4].

In this sense, mobile agents is an option to distribute and scale the network management. These agents move to the place where data are stored and select information the user wants by using intelligence; saving bandwidth, time, and money. Mobile agents decentralize processing and control, and, as a consequence, reduce the traffic around the management station, turn asynchronous the communication between the agent and the manager (useful when there are unreliable or lossy links), distribute processing load, and increase the flexibility of the management agents' behavior.

Research activities related to mobile code in network management are recent [5]. Magedanz et al. [6] have been one of the first researchers to present issues for the deployment of mobile agents in telecommunication and network management. Labetoulle et al. [7] present a state-of-the-art of intelligent agents in network management, from the management by delegation paradigm [8] to the mobile agent one, while Karmouch and Pham [5] describe the current use of mobile agents in telecommunications and network management. Management by Delegation (MbD) [8] has been the first paradigm to address decentralization and automation of management tasks by dynamically delegating management functions to agents. Pagurek et al. [9] implement a code mobility infrastructure and a suite of simple tools that interact with agents located on network components. Puliafito et al. [10] use mobile agents to collect information about the state of the network and to perform a micro-management of network devices through multiple variables. Morin et al.[11] implement a management architecture that consists of a set of servers and managers communicating with SNMP agents located on the managed network elements.

The performance of mobile agents in network management is also being investigated. Geihs et al. [12] and El-Darieby and Bieszcad [13] provide simple quantitative evaluation of mobile agent and client-server approaches. Baldi et al. [3] evaluate the tradeoffs of mobile code design paradigms in network management applications by developing a quantitative model that provides the bandwidth used by traditional and mobile code design of management functionalities. Gavalas et al. [14] analyze bandwidth utilization for the SNMP and the mobile agent. Experimental implementation results are presented in terms of bandwidth consumption and response time to obtain an aggregation of multiple variables. Sahai and Morin [11] perform measurements of bandwidth utilization of mobile agent and client-server applications on an Ethernet LAN. They also present a single case comparison of response time for the mobile agent and the client-server. Rubinstein and Duarte [15, 16] simulate management tasks performed by mobile agents and SNMP ones, comparing both the approaches, on a topology that consists of a LAN of managed network elements connected to the management station by a bottleneck link.

In this paper, we assess the performance improvement obtained by using multiple mobile agents. The network is partitioned into several domains and one mobile agent is assigned to each domain. Transit-stub topologies, which are similar in shape to the Internet, are used in the simulations. Response time results show that the management performance increases when using multiple mobile agents.

This paper is organized as follows. Section 2 discusses the use of mobile agents in network management. Section 3 reports simulation results concerning the applicability of mobile agents in management tasks. At last, concluding remarks are presented in Section 4.

2. NETWORK MANAGEMENT USING MOBILE AGENTS

Mobile agents are created to perform their tasks in different network computers. A mobile agent can interrupt its execution and migrate from a machine to another, carrying data about its state that include information obtained from previous task executions. As the number of visited nodes grows, mobile agent size also increases, turning migration harder. One possible solution to this problem is to visit a fixed number of nodes, return to the agent home or send all data to it (reducing mobile agent size), and start the task again on the remaining nodes [15]. Another solution is to use multiple mobile agents to manage a partitioned network (Section 3). The initial size of a mobile agent also affects agent performance since the larger the size, the more difficult the migration [17]. This size depends on the task to be carried out and on the language used to implement it.

In the SNMP, network management does not scale when the size or the complexity of the network increases because of the centralized processing and control. Mobile agents can be used in order to solve this problem and it is important to investigate when they improve management efficiency.

2.1 Simulation Model

The applicability of mobile agents in carrying out network management tasks is assessed by comparing mobile agent performance with the SNMP one.

The Network Simulator (NS) is used in these simulations [18]. This discrete-event simulator provides several implemented protocols and mechanisms to simulate computer networks with node and link abstractions. Some UDP modules of the NS have had to be modified in order to effectively receive a packet, sending it to the next layer.

All simulations use transit-stub topologies, which are similar in shape to the Internet. The GT-ITM [19] generates random topologies, on which each node represents switches or routers and edges represent forwarding paths between switches (Figure 1).

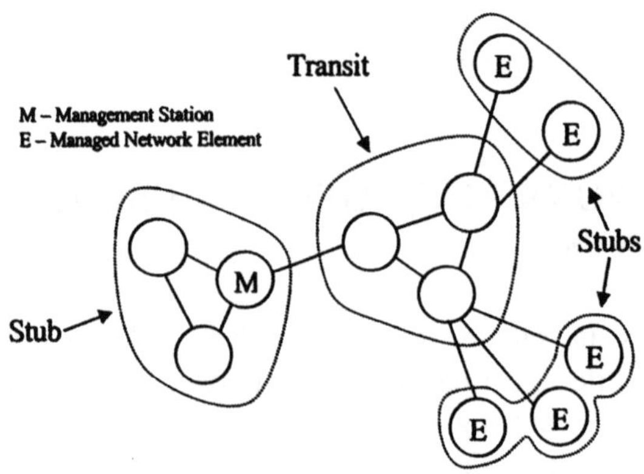

Figure 1. Network management on a transit-stub topology

Each routing domain in the Internet can be classified as either a stub domain or a transit domain. A domain is a stub domain if the path connecting any two nodes u and v goes through that domain only if either u or v is in that domain. Transit domains do not have this restriction. The purpose of transit domains is to interconnect stub domains efficiently.

A transit domain comprises a set of backbone nodes, which are typically fairly well connected to each other. In a transit domain, each backbone node also connects to a number of stub domains, via gateway nodes in the stubs.

We consider that the management station belongs to a node of a stub domain and managed network elements are located in other stub domains (Figure 1). We use the same input parameters to generate three random 272-

node topologies. The management station controls groups of 16 network elements, which is the number of nodes of a stub domain.

The simulation model assumes that links and nodes have no load, links are error-free, and processing time in application layer is not taken into account. UDP is used in all simulations and UDP and IP headers are considered. The Maximum Transmission Unit (MTU) used for all links is 1500 bytes, therefore, there is no fragmentation of SNMP messages since they are small. Every request or response of a variable is sent on a different message and the considered performance parameters are bandwidth consumption on the links that connect the management station to the backbone nodes and response time in retrieving management variables.

SNMP sends requests to all elements to be managed (one after receiving the response from the other); the mobile agent goes to an element to be managed, gathers the variable, and visits all other elements, from a stub domain to another. After finishing, the mobile agent returns to the management station.

3. SIMULATION RESULTS

First, we evaluate the effect of the number of managed network elements on response time. We assume that each request or response is 50 bytes long, all links have a 2 Mbps data rate and latency links are about a few milliseconds, and the initial size of the mobile agent is 5 kbytes.

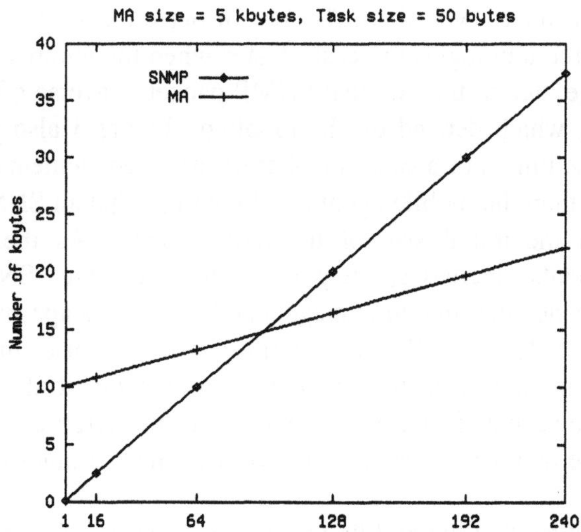

Figure 2. Bandwidth consumption per number of managed elements

For a small number of managed network elements, the SNMP uses fewer bytes than the mobile agent to perform the task (Figure 2). Nevertheless, as the number of managed elements increases, the overhead associated with several retrievals (PDU GetRequest, UDP, and IP's headers) turns mobile agent utilization more suitable.

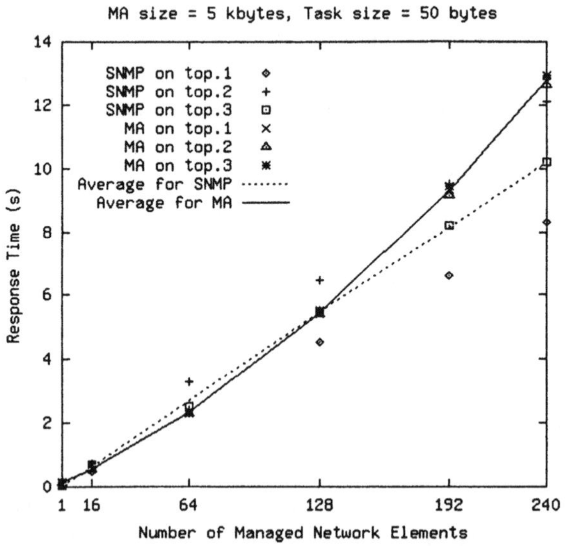

Figure 3. Response time per number of managed elements

Figure 3 shows that the mobile agent's behavior does not change with the topology, but for the SNMP, the difference of performance among response times for the three topologies increases faster when the number of managed elements is large, due to the fact that SNMP packets traverse oftentimes the backbone links, which depend on the topology. Figure 3 also presents the average response time. For a small number of managed elements, the SNMP performs better than the mobile agent due to the fact that an SNMP message is smaller than the initial size of the mobile agent. As the number of managed elements increases, response time for the SNMP grows proportionally, since the time to manage a stub domain is approximately the same for all stub domains. For the mobile agent, response time increases faster when the number of managed elements grows, due to the incremental size of the mobile agent. Therefore, the mobile agent performs better that the SNMP when the number of managed network elements ranges between two limits.

It has been show that the mobile agent size increases with the number of visited nodes, and, as a consequence, the migration becomes difficult. Moreover, when the number of elements to be managed is large, the response time for the mobile agent will also be high. In order to improve the management performance, we divide the network into several domains and assign one mobile agent to each domain. The number of mobile agents used in the simulations is 1, 2, 4, and 8. Different initial sizes of the mobile agents (S) and task sizes (T) are used, and one of the three topologies is considered.

In order to also analyze the effect of the task on the management performance, two different tasks are used: task t_1 is the same task used in the previous experiment and task t_2 is associated with a 150-byte-long variable.

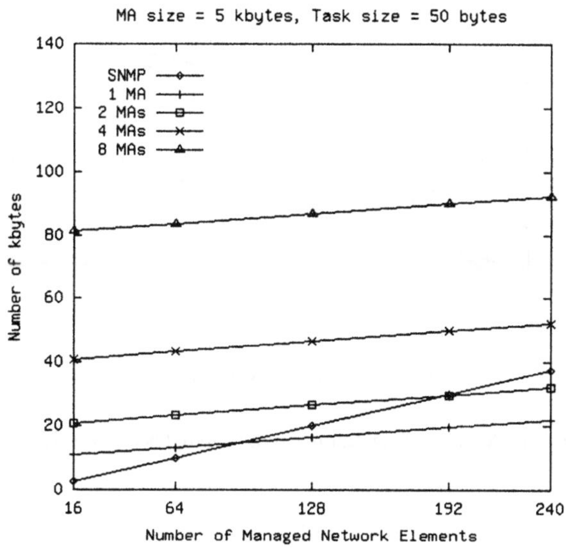

Figure 4. Bandwidth consumption for S = 5 kbytes and T = 50 bytes

Figures 4 and 5 show the bandwidth consumption per number of mobile agents for S = 5 kbytes and T = 50 and 150 bytes. The SNMP and the mobile agents are both influenced by the task to be performed. When the number of managed elements is 240, bandwidth consumption for the SNMP is 37.44 and 85.44 kbytes, respectively for T = 50 and 150 bytes; for one mobile agent, bandwidth consumption is 22.05 and 46.15 kbytes, for T = 50 and 150 bytes and 240 managed elements. When we increase the number of mobile agents, the bandwidth consumption also increases, since all mobile agents return to the management station after finishing their tasks. For example, in Figure 4, when the number of managed network elements is 240 and the number of mobile agents increases from 1 to 8, the bandwidth consumption increases from 22.05 to 92.4 kbytes.

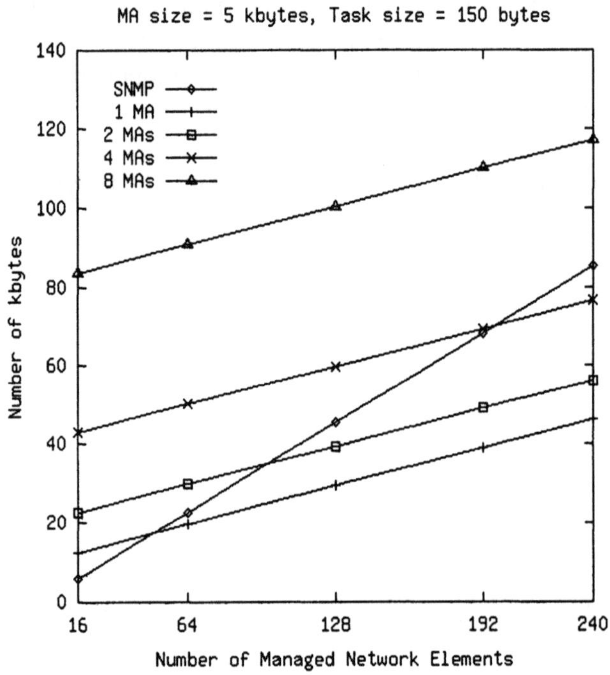

Figure 5. Bandwidth consumption for S = 5 kbytes and T = 150 bytes

For the response time, as depicted in Figures 6 and 7, the mobile agents and the SNMP also have different behaviors. For the SNMP, the task does not have a great influence due to the fact that the number of bytes exchanged between the management station and the managed network elements is small; but for the mobile agents, as the number of bytes exchanged increases, the response time also grows since the mobile agents will move with higher difficulty. For the mobile agents, as the number of managed network elements increases, the performance difference among response times for the different number of mobile agents increases faster (Figures 6 and 7). For example, when T = 50 bytes (Figure 6) and the number of managed network elements doubles from 64 to 128, the response time for 1 and 8 mobile agents goes from 2.3209 to 5.43483 s, and from 0.518067 to 0.778 s, respectively.

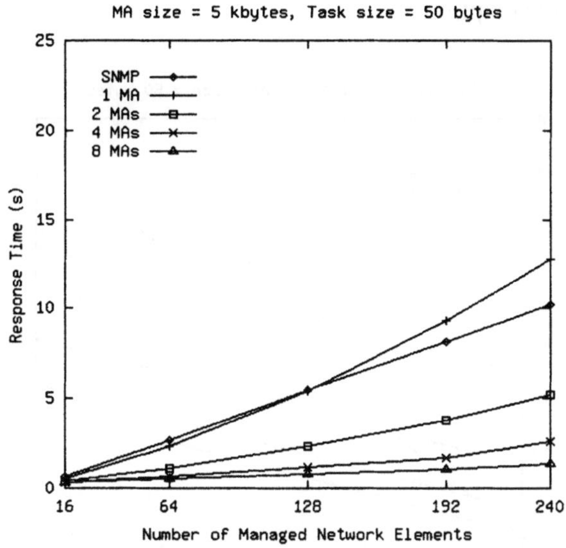

Figure 6. Response time for S = 5 kbytes and T = 50 bytes

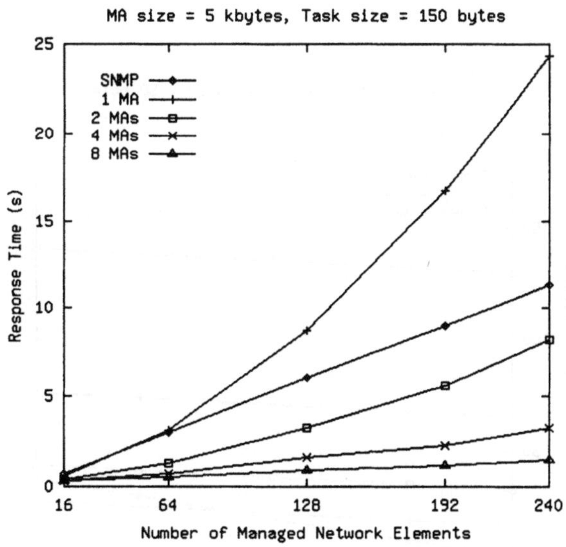

Figure 7. Response time for S = 5 kbytes and T = 150 bytes

Two different initial sizes of mobile agents are used: 1 and 8 kbytes. Figures 8 and 9 show the bandwidth consumption per number of mobile agents for T = 50 bytes and S = 1 and 8 kbytes. Comparing Figures 8 and 9, we conclude that the initial size of the mobile agent has a great influence on the bandwidth consumption.

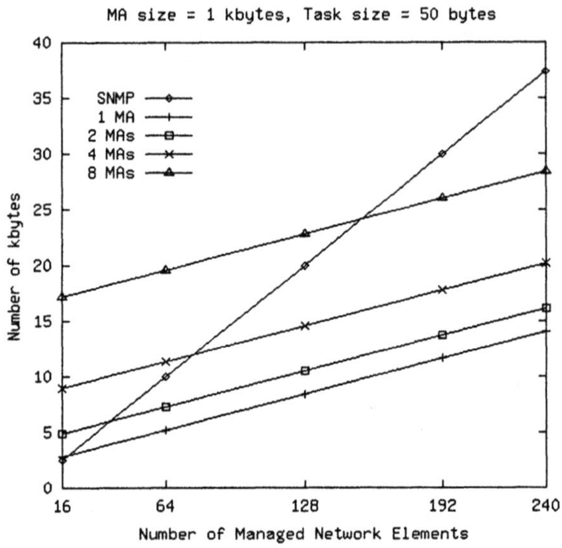

Figure 8. Bandwidth consumption for S = 1 kbytes and T = 50 bytes

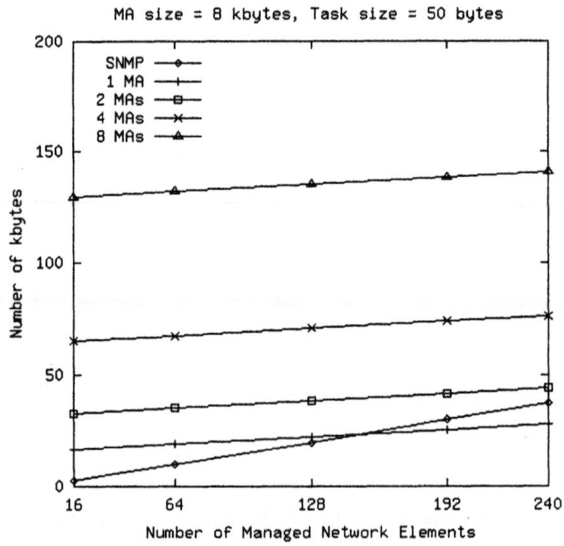

Figure 9. Bandwidth consumption for S = 8 kbytes and T = 50 bytes

Figures 10 and 11 show the response time for different mobile agent sizes. For example, when the number of managed elements is 240 and the initial size of each mobile agent goes from 1 to 8 kbytes, the response time for 1 and 8 mobile agents goes from 8.9154 to 15.6741 s, and from 0.614267 to 1.89 s, respectively.

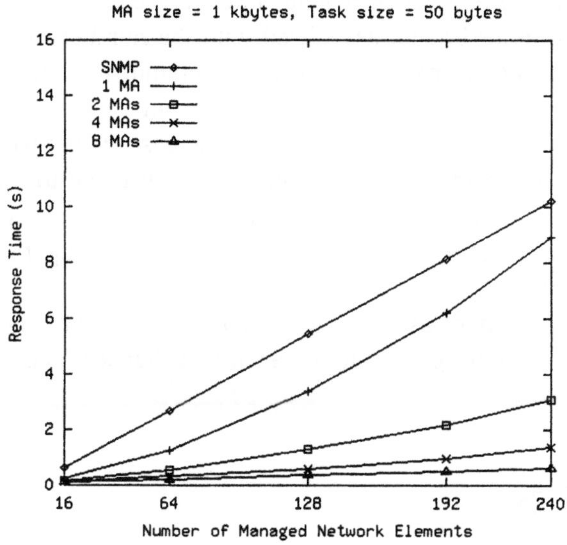

Figure 10. Response time for S = 1 kbytes and T = 50 bytes

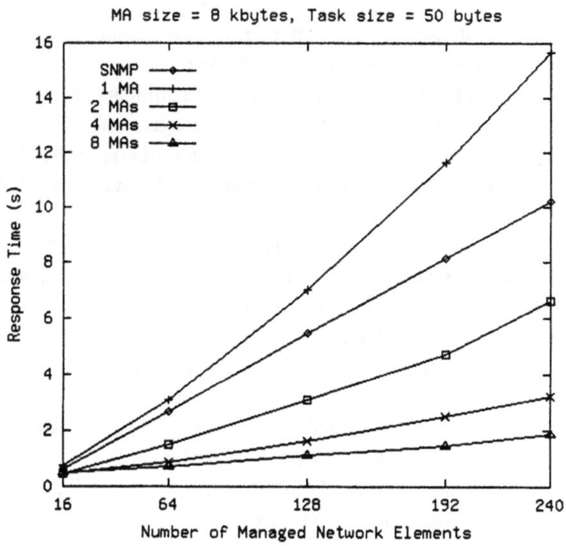

Figure 11. Response time for S = 8 kbytes and T = 50 bytes

4. CONCLUSION

This paper has analyzed tradeoffs of mobile agents in network management tasks. The performance of mobile agents has been compared with the SNMP one in several simulations.

Response time results show that the mobile agent performs better than the SNMP when the number of managed network elements ranges between two limits: an inferior bound, associated with the number of messages that pass through backbone links, and a superior bound, related to incremental size of mobile agent, which turns migration difficult.

Moreover, the management performance increases when we use multiple mobile agents. For an initial size of a mobile agent equal to 5 kbytes and a 50-byte-long task, response times for gathering 240 variables decrease by 59, 80, and 90%, for 2, 4, and 8 mobile agents, when comparing with using one mobile agent.

Network management using mobile agents seems to be a good approach and the use of multiple mobile agents enhances the management performance.

REFERENCES

[1] W. Stallings, "SNMP and SNMPv2: The infrastructure for network management", IEEE Communications Magazine, vol. 36, no. 3, pp. 37-43, Mar. 1998.

[2] Y. Yemini, "The OSI network management model", IEEE Communications Magazine, vol. 31, no. 5, pp. 20-29, May 1993.

[3] M. Baldi and G. P. Picco, "Evaluating the tradeoffs of mobile code design paradigms in network management applications", 20th International Conference on Software Engineering (ICSE'98), Kyoto, Japan, pp. 146-155, Apr. 1998.

[4] K. Meyer, M. Erlinger, J. Betser, C. Sunshine, G. Goldszmidt and Y. Yemini, "Decentralizing control and intelligence in network management", 4th International Symposium on Integrated Network Management, Santa Barbara, USA, May 1995.

[5] V. A. Pham and A. Karmouch, "Mobile software agents: An overview", IEEE Communications Magazine, vol. 36, no. 7, pp. 26-37, July 1998.

[6] T. Magedanz and T. Eckardt, "Mobile software agents: A new paradigm for telecommunications management", IEEE/IFIP Network Operations and Management Symposium (NOMS'96), Kyoto, Japan, Apr. 1996.

[7] M. Cheikhrouhou, P. Conti and J. Labetoulle, "Intelligent agents in network management: a state-of-the-art", Networking and Information Systems Journal, vol. 1, no. 1, pp. 9-38, 1998.

[8] G. Goldszmidt and Y. Yemini, "Delegated agents for network management", IEEE Communications Magazine, vol. 36, no. 3, pp. 66-70, Mar. 1998.

[9] A. Bieszczad, B. Pagurek and T. White, "Mobile agents for network management", IEEE Communications Surveys, vol. 1, no. 1, Sept. 1998.

[10] A. Puliafito and O. Tomarchio, "Using mobile agents to implement flexible network management strategies", Computer Communications, vol. 23, no. 8, pp. 708-719, Apr. 2000.

[11] A. Sahai and C. Morin, "Towards distributed and dynamic network management", IEEE/IFIP Network Operations and Management Symposium (NOMS'98), New Orleans, USA, Feb. 1998.

[12] M. Zapf, K. Herrmann and K. Geihs, "Decentralized SNMP management with mobile agents", Sixth IFIP/IEEE International Symposium on Integrated Network Management (IM'99), Boston, USA, pp. 623-635, May 1999.

[13] M. El-Darieby and A. Bieszczad, "Intelligent mobile agents: Towards network fault management automation", Sixth IFIP/IEEE International Symposium on Integrated Network Management (IM'99), Boston, USA, pp. 610-622, May 1999.

[14] D. Gavalas, D. Greenwood, M. Ghanbari and M. O'Mahony, "Using mobile agents for distributed network performance management", 3rd International Workshop on Intelligent Agents for Telecommunication Applications (IATA'99), Stockholm, Sweden, Aug. 1999.

[15] M. G. Rubinstein and O. C. M. B. Duarte, "Evaluating tradeoffs of mobile agents in network management", Networking and Information Systems Journal, vol. 2, no. 2, pp. 237-252, 1999.

[16] M. G. Rubinstein and O. C. M. B. Duarte, "Analyzing mobile agent scalability in network management", IEEE Latin American Network Operations and Management Symposium (LANOMS'99), Rio de Janeiro, Brazil, pp. 64-74, Dec. 1999.

[17] M. G. Rubinstein and O. C. M. B. Duarte, "Evaluating the performance of mobile agents in network management", IEEE Global Telecommunications Conference (GLOBECOM'99), Rio de Janeiro, Brazil, pp. 386-390, Dec. 1999.

[18] K. Fall and K. Varadhan, "NS Notes and Documentation", tech. rep., The VINT Project, Jan. 1999.

[19] E. W. Zegura, K. L. Calvert and M. J. Donahoo, "A quantitative comparison of graph-based models for internet topology", IEEE/ACM Transactions on Networking, vol. 5, no. 6, pp. 770-783, Dec. 1997.

Chapter 19

Hierarchical Rerouting Model for Fault Tolerance in Multi-Network Environment

Won-Kyu Hong, Dong-Il Kim, Seong-Sook Yoon, Seong-Ik Hong, Mun-Jo Jung
Telecomminications Network Lab., R&D Group, Korea Telecom
463-1, Junmin-dong, Yusung-gu, Taejeon, Korea
Phone : +82-42-870-8254 Fax : +82-42-870-8229
{wkhong,dikim,pseudo,yeolin,mjjung}@kt.co.kr

Key words: Hierarchical Network Management, Rerouting, Fault Tolerance, CORBA

Abstract: ATM virtual path has recently been paying attention to the effective deployment of IP over ATM. The effective IP service provisioning depends fully on the reliability and optimality of ATM VP layer network because the relationship between IP and ATM VP networks is client and server. In order to guarantee the network survivability of ATM VP layer network and to provide the reliable IP service provisioning, this paper proposes a hierarchical rerouting model in multi-network environment of ATM VP and IP networks. In addition, this paper proposes the alarm correlation and localization model in the hierarchical transport network. And this paper describes the COBRA-based implementation model and its experience.

1. INTRODUCTION

Due to the rapidly growing demand for multimedia information transfer across the communication networks, the need for reliable communication service become more important. The potential effects of communication network failures have been demonstrated by several publicized studies, showing the need for survivable networks that are robust to failures [2,3]. A variety of network failures typically result from a failure by accidental cable cuts, hardware malfunctions, software errors, natural disasters, and human errors. Typically, failures can be classified into physical and software failures. Considering the failure by a fiber cable cut in a self-healing ring in a metropolitan area telecommunication network, we can restore it within several milliseconds by changing the direction of flow with other available backup links. However, using a ring approach in a national or an international network is likely to be cost-prohibitive. Because the spare capacity of the backup link provided for such restoration is unused during normal network operation and so we can make no profit by using the resource while being idle. Therefore, it is not cost-effective to provide for transmission level restoration in many networks.

The goal of fault recovery is to restore a large fraction of connections in usually less than two seconds. The restoration is performed at the lower levels of the transport hierarchy, such as the SONET/SDH transmission path layer, and is based on self-healing rings, spare backup trunks, and careful dimensioning of the network [6]. In an ATM based network, most restoration is performed at the Virtual Path (VP) layer, and consequently it provides the same QoS to all connections of the VP. There are three important issues related with restoration. One is how to correlate the alarms that may be generated by different network elements due to a failure, another is a fault isolation and localization, and the other is the time required to restore a failure by taking appropriate actions and the transient network behaviour.

On the other hand, layering concept is recently applied for the effective deployment of IP network and ATM network in terms of IP over ATM. The survivability of ATM network as a backbone of IP data network effects the reliable IP service provisioning because the relationship between ATM VP and IP layer networks is server and client. In order to guarantee the network survivability of ATM VP layer network and to provide the reliable IP service provisioning, this paper proposes a multi-layer network model that can commonly be applicable for IP network based on ATM network. This paper also proposes a hierarchical rerouting model based on the multi-layer network model that is very cost-effective because it does not need any extra capacities such as self-healing ring, spare trunks, etc. In addition, this paper proposes the alarm correlation and localization model in the hierarchical transport network. And this paper describes the COBRA-based implementation model and its experience.

2. HIERARCHICAL MULTI-NETWORK MODEL

In order to take the advantages of the capability of QoS guarantee of ATM and the capacity of diverse multimedia traffic delivery of IP network, we construct IP network over ATM backbone using the ITU-T G.805

layering concept [7]. ITU-T G.805 defines the ATM transport network architecture with the layering and partitioning concepts. From the perspective of layering concept, ATM VP network takes the role of server layer network and IP network takes the role of client layer network. All the Routers in IP network are interconnected with ATM VPC provided by ATM VP layer network and all the ATM VC switches in VC layer network are interconnected with ATM VPC provided by ATM VP layer network. Each layer network is designed in line with the ITU-T G.805 partitioning concept [1], ATMForum element view [7] and network view [8] as shown in Figure 1. Layer network partitioning aims to provide the network scalability and manageability and to promote the fault isolation and localization within the partitioned domain. Network Partitioning is normally done as the TMN functional layering concepts: Network Management Layer, Element Management Layer and Network Element.

Layer network is a topological component that includes both transport entities and transport processing functions that describe the generation, transport and termination of the particular characteristic information. Subnetwork (SNW) is a topological component that is used to define routing and rerouting domains of the specific characteristic information.

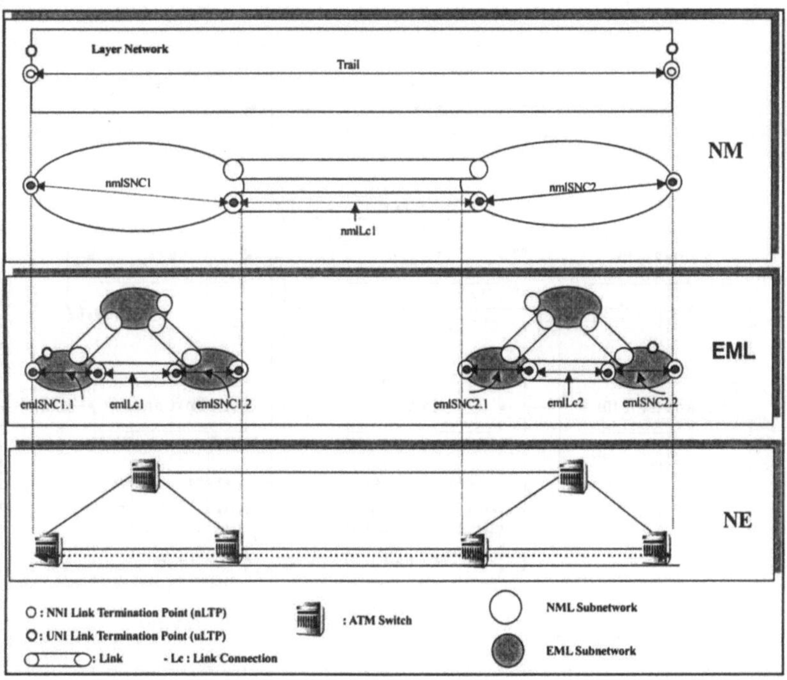

Figure 1. Hierarchical Network Model

Link is a topological component that describes a fixed relationship between one subnetwork and another subnetwork. Link connection (LinkConn) represents a transport entity that transfers information between nLtps across a link. UNI Link Termination Point (uLtp) is a reference point having the adaptation function between client and server layer network. NNI Link Termination Point (nLtp) is a reference point terminating link connection. Trail is a transport entity that is responsible for the transfer of specific characteristic information. Subnetwork connection is a transport entity that transfers information across a subnetwork and is formed by the association (or binding) of Ltps on boundary of the subnetwork. Connection Termination Point (ctp) is a reference point that is consisted of a pair of co-located bi-directional connections. Routeprofile is a transport entity that represents the route information of NML Subnetwork level. Subrouteprofile is a transport entity that represents the route information of EML Subnetwork level. Hierarchical transport network is described in terms of Rumbaugh's object modeling notation [9] as shown in Figure 2. It shows the object relationship between topological objects and connectivity objects.

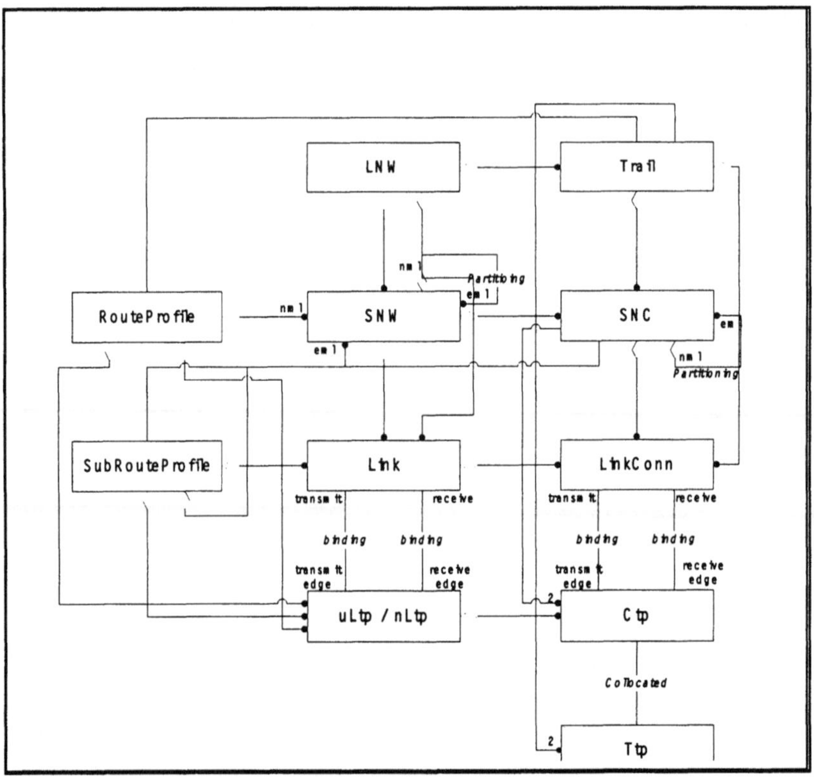

Figure 2. Hierarchical Transport Network Model

3. ALARM CORRELATION MODEL

Alarm correlation is generally made on the rule by which several alarms are narrowed from a mass of problems to a root cause and side effects. We define our own *Alarm Correlation Model (ACM)* taking into account the effective rerouting in the hierarchical transport network as our rules to identify the *Avoidance List (AL)* and to determine *Rerouting List (RL)* according to the fault location and types. The avoidance list represents the set of topological components that should be excluded in the rerouted path and the rerouting list represents the set of connectivity components that should be rerouted in the process of restoration.

We define only four kinds of alarm types for simplicity – UNI LTP (uLtp), NNI LTP (nLtp), SNW, and SNC. However we do not try to reroute in the case of uLtp alarm because there is no any alternative path. So we deal with only nLtp, SNW and SNC alarm as shown in Figure 3.

Our alarm correlation model is purely rule-based one as below:

In the case of *SNC fault* at Figure 3, the avoidance lists are the adjacent links (Link1 and Link2) connected with the nLtps (nLtp2 and nLtp3) that the fault SNC (SNC2) is terminated.

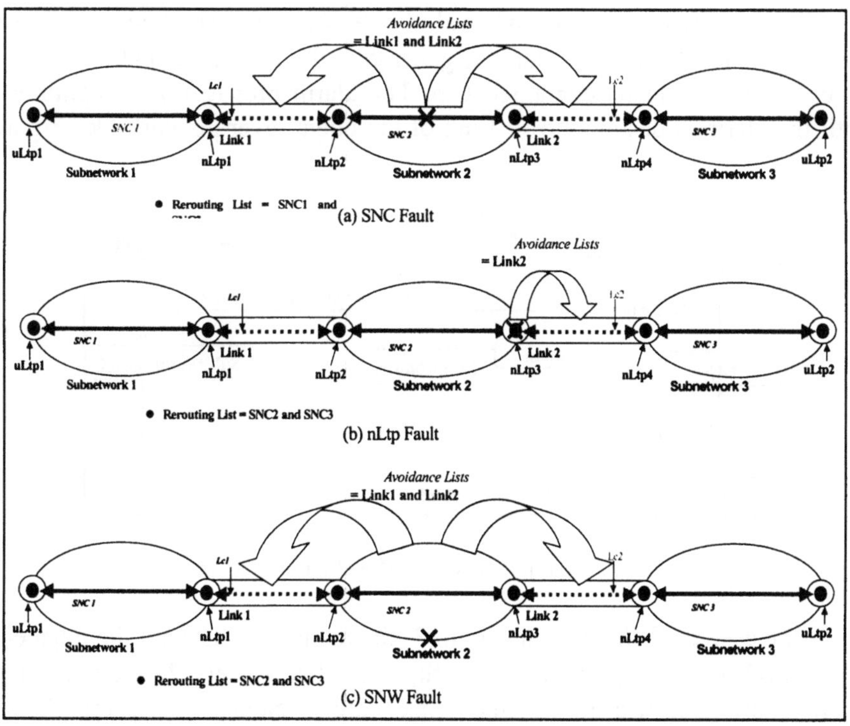

Figure 3. Alarm Correlation Model

The rerouting lists are the SNC1 terminated at the nLtp1 and the SNC3 terminated at the nLtp4 because the Link1 and Link2 are avoidance lists. When SNC alarm is generated, the SNC propagates alarm to its superior SNC and its terminating points of Ctps. The alarm is continuously propagated until the alarm is destined to IP Link of the client layer network like Figure 4.

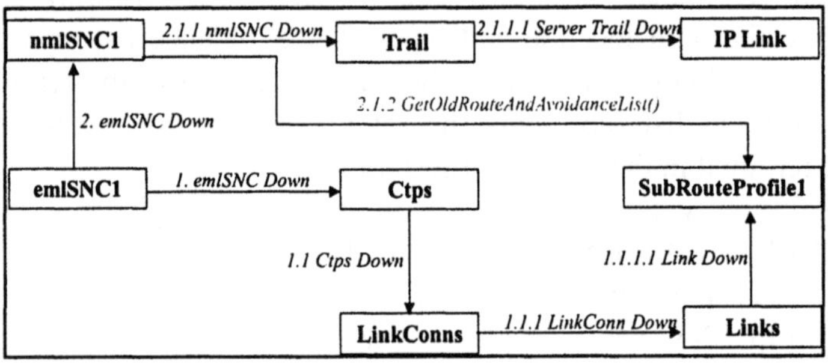

Figure 4. Alarm Propagation Model for SNC Alarm

In the case of *nLtp fault* at Figure 3, the avoidance list is the link (Link2) connected with the failed nLtp (nLtp3). Because Link2 is avoidance list, the SNC3 terminated at the nLtp4 and the SNC2 terminated at nLtp3 are rerouting lists. When nLtp alarm is generated, the nLtp propagates alarm to its bound Link and its contained Ctp. The alarm propagation is continued until the alarm is destined to IP Link of the client layer network like Figure 5.

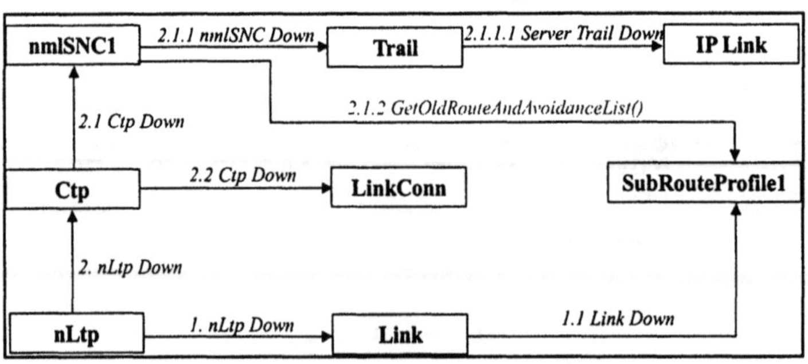

Figure 5. Alarm Propagation Model for nLtp Alarm

In the case of *SNW fault* at Figure 3, the avoidance lists are the adjacent links (link1 and link3) that belong to the failed SNW (SNW2) and the failed SNW itself. Because Link1 and Link2 are avoidance lists, the SNC1 terminated at the nLtp1 and the SNC3 terminated at the nLtp4 are rerouting lists. When SNW alarm is generated, the SNW propagates alarm to the Ltps and the SNCs within the fault SNW. The alarm propagation is continued until the alarm is destined to IP Link of the client layer network like Figure 6.

In addition to identify the avoidance list and rerouting list, there is another difficult problem that is how we can determine and minimize the portion of restoration affected by a certain of alarm. It means the identification of rerouting scope. Fortunately, there is a useful concept of ITU-T G.805 partitioning [1] to determine the rerouting scope. A layer network can be partitioned into subnetworks according to the administrative domain or management boundary. Also, the partitioned subnetworks are recursively partitioned until the lastly partitioned subnetwork is corresponding to a switching fabric as shown in Figure 1.

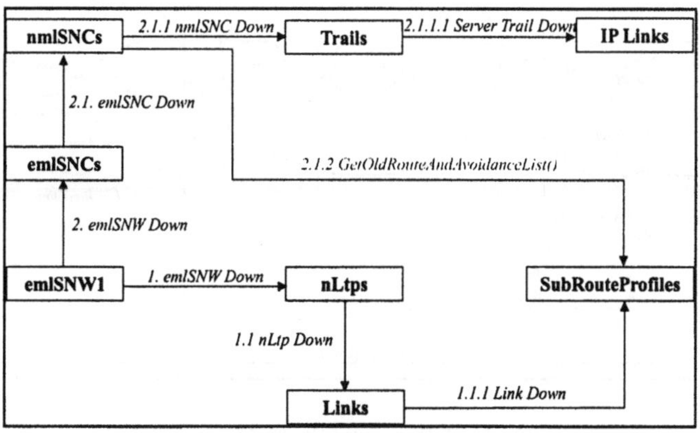

Figure 6. Alarm Propagation Model for SNW Alarm

Therefore the partitioned subnetwork itself can be the rerouting scope. The identification of avoidance list and routing list for rerouting is the role of NML function. There is another important concept for localizing rerouting scope. Rerouting scope in hierarchical transport network is determined by the location of alarm in hierarchical transport network like Figure 7. There are two kinds of rerouting scopes – subnetwork and layer network level. If an alarm is occurred at any object within a NML subnetwork (nmlSNW), its rerouting scope is the NML subnetwork. If an alarm is occurred at nLtps at the boundary of NML subnetwork, its rerouting scope is the layer network.

For example, in the first case of fault at Figure 7, its rerouting scope is nmlSNW1 and routed connection is nmlSNC1. If the restoration of nmlSNC1 fails within the nmlSNW1, its rerouting scope is expanded to layer network of Trail. In the second case of fault at Figure 7, the alarm is occurred at the nLtp of the boundary nmlSNW2 and its avoidance list is the Link between nmlSNW2 and nmlSNW3. The avoidance list of Link is managed by layer network and its rerouting scope is layer network of Trail.

There is another kind of alarm that cannot be restored in this model and any other models. It is the alarm of uLtp or nLtp that is the termination point of Trail like the third case of fault at Figure 7. Because we confine the end-to-end connection management scope in the UNI-to-UNI or NNI-to-NNI excepting the UNI Link, there are no more alternatives to restore such kinds of faults as the uLtp and nLtp containing the terminating trail termination point.

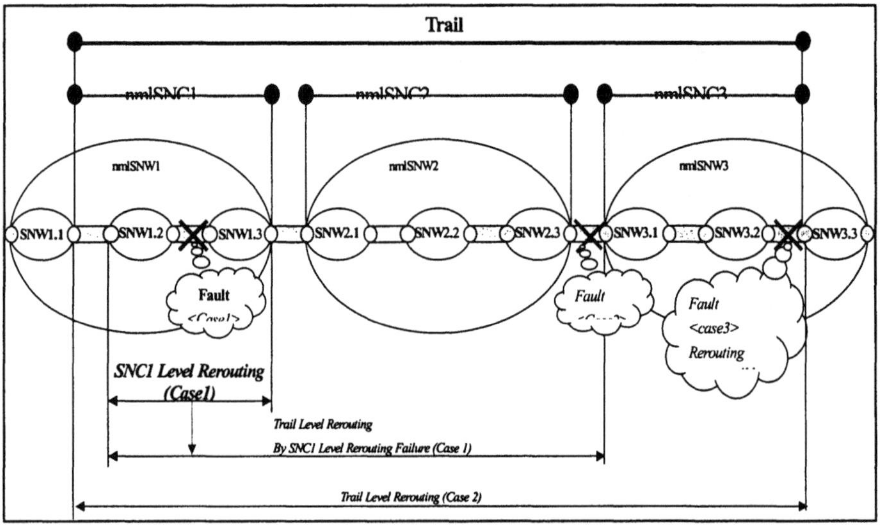

Figure 7. Rerouting Scope Model

4. HIERARCHICAL REROUTING MODEL

Our rerouting procedure is composed of five steps: (1) alarm propagation, (2) determination of avoidance list, (3) identification of rerouting list, (4) finding alternative route avoiding the avoidance list, and (5) restoration by SNC manipulation. Rerouting function is the role of the Network Management Layer (NML) function of TMN functional architecture. Therefore, there is not any rerouting function in the Element Management Layer (EML) but notification function of alarm occurrence to Network Management Layer (NML). Figure 8 shows the hierarchical rerouting model, which an end-to-end connection of Trail is composed of two nmlSNCs of nmlSNC1 and nmlSNC2. And each nmlSNCs are composed of two emlSNCs. This paper describes the rerouting procedure when the nLtp is down as shown in Figure 8 in line with the hierarchical rerouting procedure as shown in Figure 9.

Alarm Generation – A NNI Ltp (nLtp) down is notified to network management system.

Alarm Propagation - Alarm is propagated according to our alarm propagation rule. In the case of nLtp down, nLtp down alarm is propagated with the alarm propagation model of nLtp alarm of Figure 5.

Determination of Avoidance List – The first rerouting scope is nmlSNC1. The nmlSNC1 determines the avoidance list with the routing information stored in the SubRouteProfile. The avoidance list is Link1.1.

Find Alternative Route – The nmlSNC1 finds the alternative route that detours the avoidance list of Link1.1. There is one alternative route traversing the Link1.2 and Link1.3.

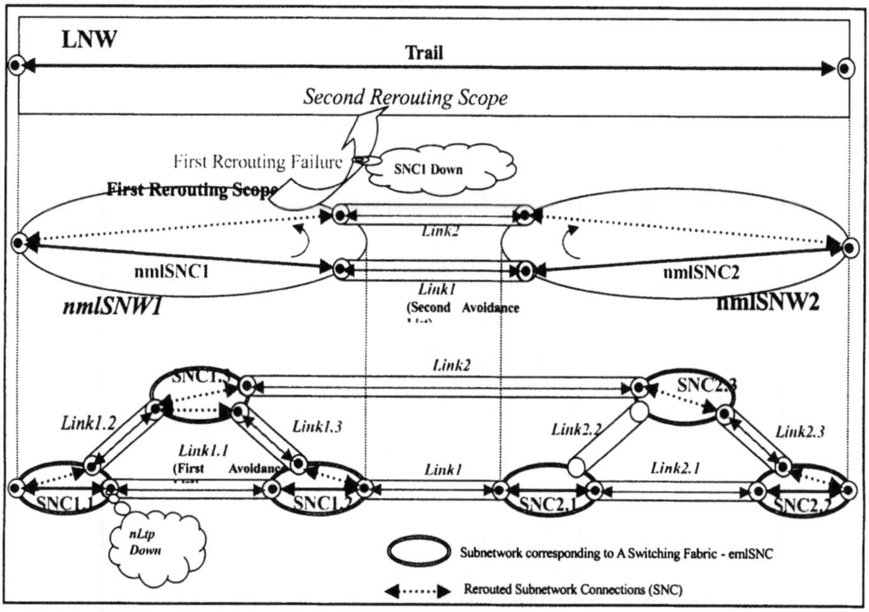

Figure 8. Hierarchical Rerouting Model

Identification of Rerouting List – As the comparison of old with alternative routes, it determines the rerouting list. Because rerouting list is the SNC that one of source or destination Ctp is changed comparing old with alternative route, the identified rerouting lists are SNC1.1 and SNC1.2.

Restoration by SNC Manipulation – There are three types of SNC manipulation for restoration: rerouting, creation, and deletion. The SNC that is not pertained in the old route but new route is newly created. The SNC that is not pertained in the alternative route but old route is deleted. With these rules, the SNC1.1 and SNC1.2 are rerouted and SNC1.3 is newly created. Whereas there is no deleted SNC because there is no SNC that is not pertained in alternative route but old route.

The restoration is done within the nmlSNC level. If the restoration of nmlSNC level failed, its rerouting scope should be expanded to the Trail level. And the avoidance list will be the Link1 and the rerouting lists will be nmlSNC1 and nmlSNC2.

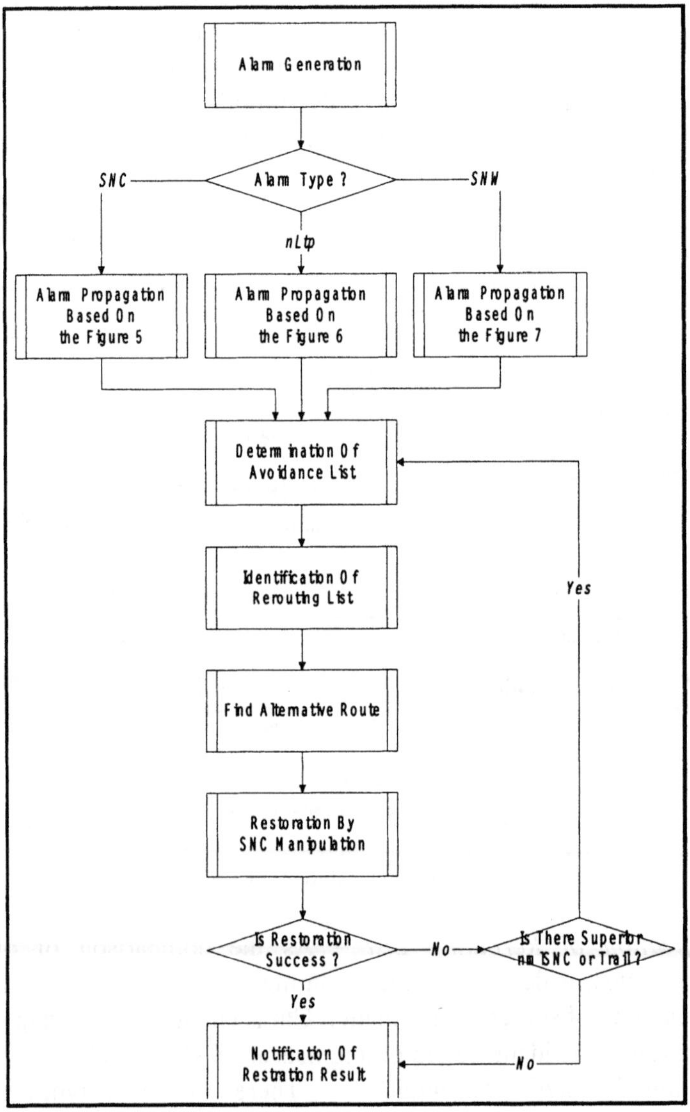

Figure 9. Hierarchical Rerouting Procedure

5. IMPLEMENTATION AND ITS EXPERIENCES

We implement the proposed hierarchical rerouting model with the CORBA platform to support distributed network management. We use the CORBA Event Service (ES) for alarm propagation. Our implementation model is shown in Figure 10.

Configuration Manger (ConfManger) takes the roles of provisioning and status monitoring of topological entities: layer network, subnetwork, uLtp, nLtp and link. Connection Manger (ConManager) takes the roles of creation, deletion and modification of connectivity entities of Trail, Subnetwork Connection, Link Connection, Trail Termination Point and Connection Termination Point. It also takes the major role of restoration by rerouting, creation and deletion of SNC in the procedure of *Restoration by SNC Manipulation.* Alarm Handler (AlarmHandler) takes the roles of alarm propagation and logging. Route Manager (RouteManager) maintains the routing tables and provides an optimal route in the case of connection setup and an alternative route in the case of rerouting. The routing and rerouting functions are solely defined in the NMS.

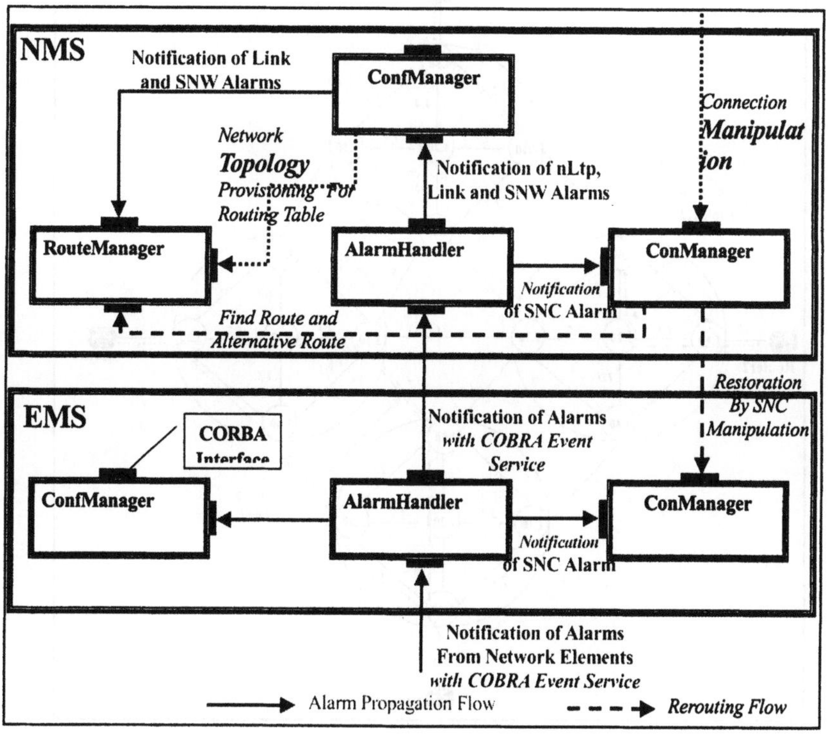

Figure 10. Implementation Model

We measure the empirical performance to validate the proposed rerouting model with the network topology like Figure 11. There are four nmlSNWs and each nmlSNW maintains the five emlSNWs. We measure the average restoration time in the two cases of the link down – one is contained within the nmlSNW, the other is contained within the layer network.

By our empirical performance analysis, we learn that the proposed hierarchical rerouting model can be applied to the large-scale ATM backbone network where each node is connected with STM-1. Figure 12 also shows the empirical VP restoration time in the case of nmlSNC and Trail level restoration with the test topology. The restoration time is measured in the case of failure of the STM-1 link that contains up to 10 VP connections. The average connection restoration time of nmlSNC level takes 1.1 seconds and that of Trail level takes 2.8 seconds including alarm propagation because of the restoration failure of nmlSNC level. The proposed rerouting model can be applicable for VP network taking the role of server layer of IP network. In the case of data network, average 2.8 seconds restoration time is reasonable.

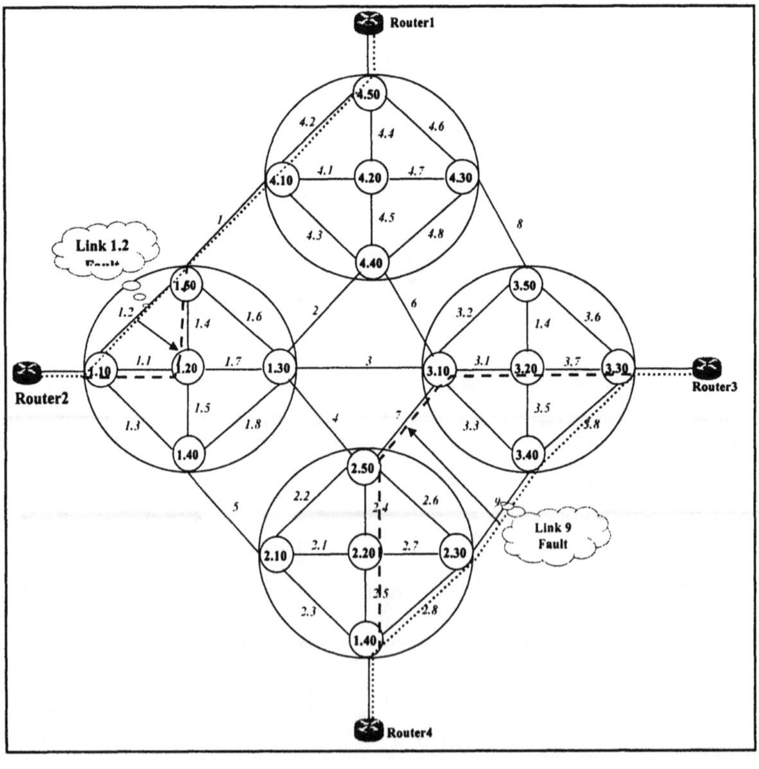

Figure 11. Network Topology for Empirical Performance Analysis

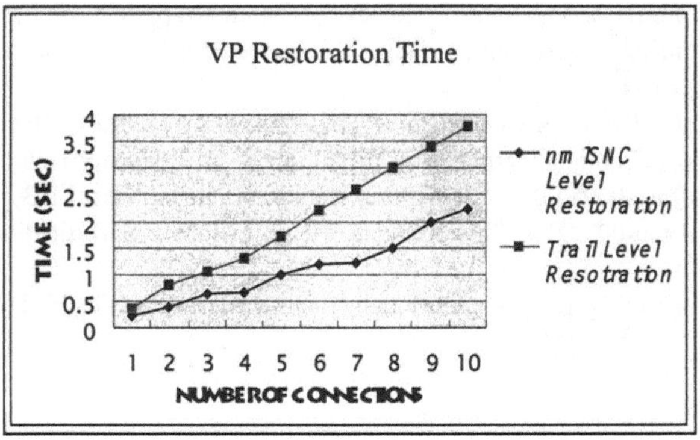

Figure 12. VP Restoration Time

6. CONCLUDING REMARKS

This paper proposes the hierarchical transport network model that can be applicable for multi-network environments. In the multi-network environment of IP and ATM VP layer networks, the reliability of VP layer network is essential for robust and efficient IP service provisioning. For reliable VP network management, this paper proposes a rule-based alarm correlation model fitted with the proposed hierarchical transport network and a hierarchical rerouting model applicable for VP layer network without any extra network facilities for restoration. With the proposed two levels of restoration – nmlSNC and Trail, we makes the restoration performance better by isolating the restoration scope within nmlSNC and increases the restoration probability by expanding the restoration scope to Trail in the case of restoration failure of nmlSNC level. Our empirical performance analysis shows that the proposed hierarchical rerouting model can be applicable for the large-scale ATM VP network with average 2.8 seconds restoration time.

REFERENCES

[1] ITU-T G.805, "Generic Functional Architecture of Transport Networks," November 19995.

[2] W. E. Falconer, "Service Assurance in Modern Telecommunications Network," IEEE Communications Magazine, Vol. 28(6), pp. 32-39, June 1990.

[3] T. H. Wu, Fiber Network Service Survivability, Artech House, Boston, Mass., 1992.

[4] Branerjea, "Simulation study of the capacity effects of dispersity routing for fault tolerant realtime channels," in Proceedings SIGCOM Symposium, pp. 194-205, August 1996.

[5] Banerjea, C. J. Parris, and D. Ferrari, "Recovery guaranteed performance service connection from single and multiple faults," in Proceedings IEEE Globecom, pp. 162-168, November 1994.
[6] J. C. McDonald, "Public network integrity – avoiding a crisis in trust," IEEE Journal on Selected Areas in Communications, vol. 12, no. 1, pp. 5-12, January 1994.
[7] ATMFroum, af-nm-0095.0001, "SNMP M4 Network Element View MIB," July 1998.
[8] ATMFroum, af-nm-0073.000, "M4 Network View CMIP MIB Specification Version 1.0," January 1997.
[9] James Rumbaugh, "Object-Oriented Modeling and Design," Prentice Hall, 1991.

Chapter 20

NetDoM: An Information System for Network Documentation Management

Alisson Sellaro, Tiago Macambira, Javam C. Machado, José Neuman de Souza
Computer Science Dept. – Federal University of Ceará, Brazil

Key words: Network Resource Documentation, Network Management, SNMP, Information Systems

Abstract: Diversification, wide spread use and heterogeneity of computer networks have demanded considerable amount of effort from researchers and vendors of network equipments. Despite these efforts, network resources documentation has apparently been neglected. This paper analyses available network documentation techniques highlighting their weak points and possible problematic approaches. Survey results about network documentation techniques used by academic and commercial communities are also presented and analyzed. Finally, information system is proposed that satisfies actual network resources documentation techniques requirements.

1. INTRODUCTION

As interconnection between computers increased in size and heterogeneity, researchers and network equipment vendors from all over the world had to unify efforts to develop tools and techniques adequate to properly manage elements connected to network. In spite of high development in areas such pro-active management systems, management protocols such as SNMP and RMON, an important area of management has been apparently neglected: network documentation.

Network documentation is defined as the act of describing computers, services, administrative procedures and occurrences in human-*readable* language in such a way that information can be retrieved for later use. Thus, good network documentation is one that describes the whole network and that is adequately updated.

In section 2 we confront our definition of network documentation with techniques used by academic and commercial communities. We will highlight weaknesses and concise points regarding each technique. In section 3 we present an applied study concerning network documentation: results of an on-line survey.

Section 4 presents our new proposed information system for management of network management documents, while section 6 describes a case study using the proposed system to document HTTP servers. In section 7 we expose our conclusion concerning network documentation techniques and the new proposed information system and suggest extensions to this study.

2. NETWORK DOCUMENTATION TECHNIQUES

There is a vast range of network documentation techniques available. HTML pages and SNMP-Based Documentation are some of the most widely used methodologies.

2.1 HTML Pages

HTML pages are the easiest way to provide well-accessible information since you can make them available to the whole Internet just by putting them on a web server directory. Figure 1 shows HTML-style documentation.

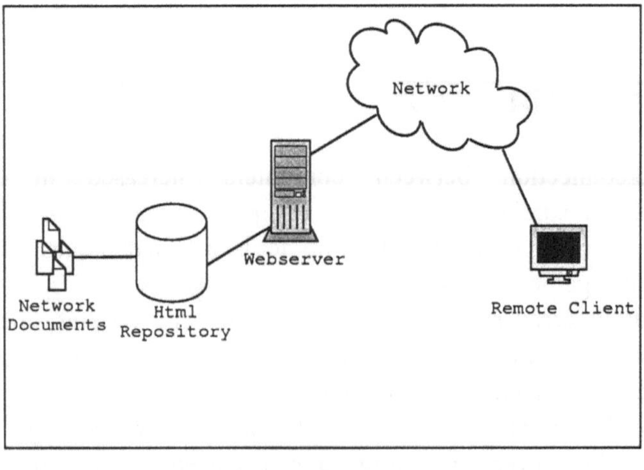

Figure 1: HTML Documentation Architecture

Different network documentation examples using HTML technique may be found in **[11, 13, 12]**.

The easy accessibility provided by HTML pages implies, generally, lack of security. Many organizations address this security problem with user-authentication control provided as features of many commercial and free HTTP servers so undesirable external users can not access sensitive organization data concerning computer network status and configuration.

Another problem with HTML-style documentation is how *to keep data up to date*. HTML pages are mostly static, human-written files. Considering the high dynamism of some managed objects, a company requiring up-to-date data would have to allocate employees specifically to rewrite HTML pages. Dynamic generation of HTML pages is a viable solution to address this problem. Powerful script languages as Perl and PHP allied with databases and web servers may be successfully used to generate simple systems to deal with updates of documents.

A good solution is to use more than one methodology for documentation, restricting the domain of HTML pages only to describe static behaviors of network such as topology maps, network management contact list etc. Figure 2 shows an example of HTML pages for documentation of static network features, the backbone map of the Brazilian National Research Network (RNP).

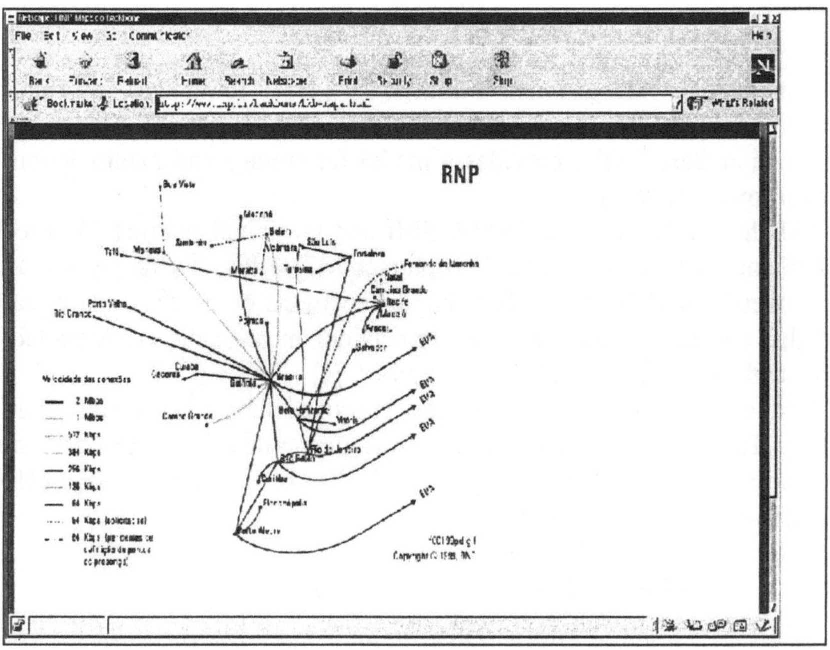

Figure 2: RNP Backbone Map

2.2 SNMP-Based Documentation

2.2.1 A Simple Introduction to SNMP

SNMP, the ``Simple Network Management Protocol" is an Internet Architecture Board (IAB) Standard proposed in RFC 1098 [4]. SNMP architecture is described in figure 3.

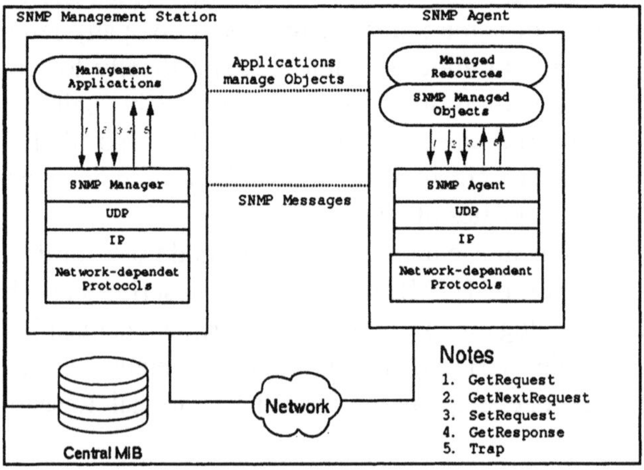

Figure 3: SNMP Architecture

Structure for Management Information (SMI) formalizes information domain subject to management under SNMP while Management Information Base (MIB) provides a model for storage and manipulation of this information. [9, 8].

IAB has designated the SNMP, SMI and the initial Internet MIB to be full "Standard Protocols" with ``Recommended" status. By doing this, IAB recommends that all IP and TCP implementations be network manageable and that the implementations that are network manageable are expected to adopt and implement SMI, MIB and SNMP. [4]

Further development of SNMP has improved functionality and security model. This new version of SNMP was denominated SNMPv2. A new framework for information management had to be designed. Thus, SMI v2 and MIB-II were developed. [5]

2.2.2 SNMP Based Network Documentation

Even though SNMP has been developed to provide a simple yet powerful management framework, a considerable amount of information it deals with is strictly dynamic. *Dynamic information* may be defined as data that vary several times in a short time period. Consider, for example, the amount of bytes a FTP server has sent to clients since its installation or the number of successful logins in a machine. This information may be considered dynamic information.

For the sake of organization and utility, to document dynamic information is not feasible, since the main objective of network documentation is to provide configuration and no status information about a given network.

A problem found in SNMP-based documentation technique lies on information accessibility. Although RMON addressed part of remote access to SNMP based tools, features as community passwords and other SNMPv2 security enhancements contribute to a very complex and not functional framework for users trying to retrieve network documents **[5]**.

3. AN ON-LINE SURVEY ABOUT DOCUMENTATION

A concise methodology to document computers networks must be formal enough to provide a high level of reliability and yet be useful. To address the exact needs and deficiencies of both academic and commercial networks, we have developed a simple survey.

3.1 Methodology

To accomplish the above-mentioned objectives we have divided our information domain into five distinct parts: number of computers managed, size of network management team, provided services, documentation tools used and institution main activity. This partitioning has allowed us to correlate all parts of information domain with documentation techniques applied in management matters.

Number of computer managed domain was discretized in four intervals: less than five, between five and twenty, between twenty one and one hundred and more than one hundred. *Size of network team* was divided in six intervals: one employee, between two and five employees, between twenty one and fifty employees, between fifty one and seventy employees and more than seventy employees.

Provided services and *documentation tools used* were multiple-choice fields. PPP/SLIP access, telnet access, POP3/IMAP, Print Spooling, FTP, NNTP, NFS or similar, NIS or similar, HTTP, SMTP and other were *provided services* options while SNMP based, MIB to document based,

HTML pages, made by myself, none and other were *documentation tools used* possible options.

Institution main activity was approached aiming to identify what specific community, academic or commercial, has bigger scarcity concerning network documentation. This domain was divided in six partitions: Internet Service Providers (ISP), university, government, commercial, research and other. ISPs were isolated studied since this kind of institution provides specifically network services.

Survey form was made available to network management community in the Web at *http://doc.mcc.ufc.br*. Call for data was sent to the following exhaustive Usenet newsgroups: comp.os.ms-windows.nt.admin.networking, comp.os.linux.networking, comp.os.ms-windows.networking.tcp-ip, comp.os.netware.connectivity, comp.dcom.net-management, comp.sys.net-computer.misc, comp.sys.amiga.networking, comp.os.linux.networking, comp.os.ms-windows.networking.misc, comp.os.os2.networking.misc, and comp.os.os2.networking.tcp-ip. We received 64 complete forms answered.

3.2 Results and Statistical Inferences

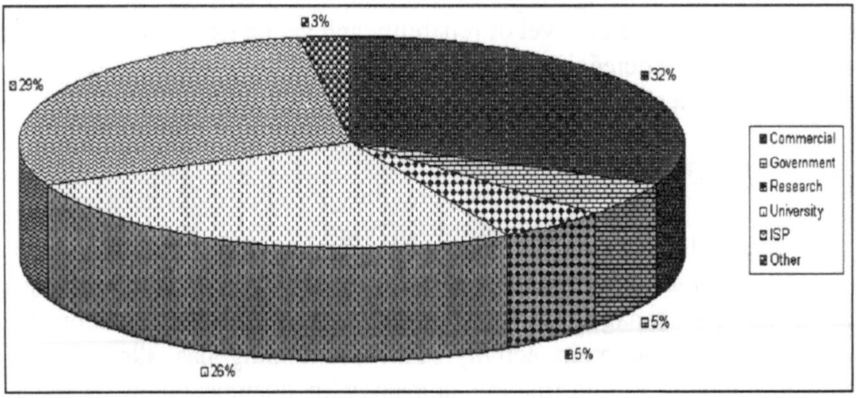

Figure 4: Institutions Distributed by Main Activity

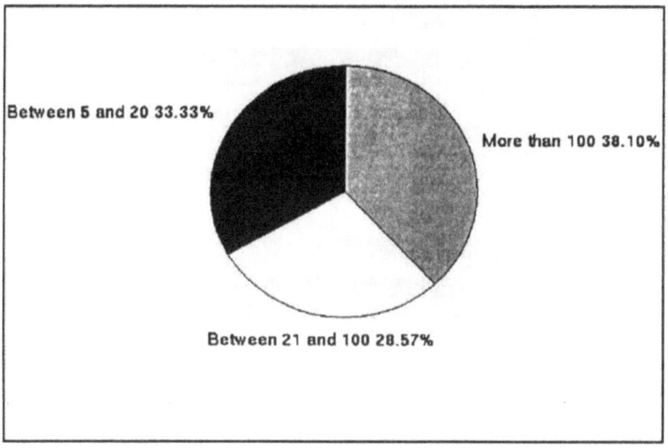

Figure 5: Number of Managed Computers

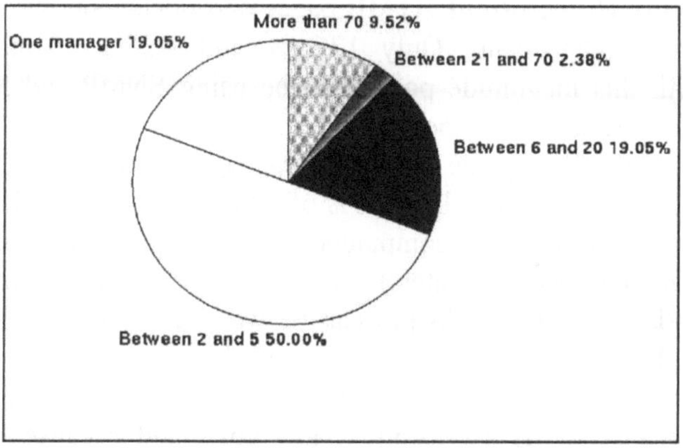

Figure 6: Network Management Team Size

3.3 Correlations

As seen in figure 5, the number of managed computers varies from five to more than one hundred, dividing this domain in only three parts. This shows that to document networks with a very small number of computers is either unnecessary or too much expensive to the network management team.

Figure 7 shows the correlation between the number of managed computers and documentation techniques used.

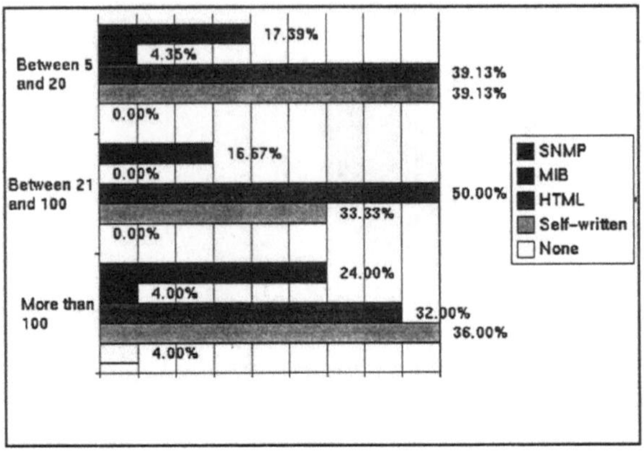

Figure 7: Number of Managed Computers X Documentation Technique Used

Small networks (between five and twenty computers) are mostly documented trough the use of HTML pages and documents written by the network management team. Only 17.39% and 4.35% of managers of networks with this magnitude pointed to be using SNMP and MIB based documentation techniques, respectively.

Medium-size and big-size networks' managers appear to use HTML documentation excessively. About 50% of medium-size networks (between twenty one and one hundred computers) and 46.67% of big-size networks (more than one hundred computers) are documented by HTML pages. Even though we admit that HTML pages can be well suited for small networks, the use of HTML pages in medium and big networks as the main documentation technique leads us to think that network managers use this technique because of its accessibility, low costs and because it allows a considerable dispersion of the management team without harming information accessibility. As highlighted in section 2.1, update problems may make network documentation a very expensive task for an institution.

SNMP-based documentation is mostly used in big networks. This is acceptable, as a viable management of networks with this magnitude requires some kind of automated tools and monitoring systems. Although section 2.2 points out that SNMP-based tools have a good performance concerning automated management SNMP-based documentation, they are not suitable because of their inherently dynamic aspect.

An impressive result found there is a high utilization of self-written documents. In all network sizes, self-written documents correspond at least to one third of all documentation techniques used. This implies a severe lack of standardization of network documents across institutions and, depending on the size of the network management team, even in the same institution.

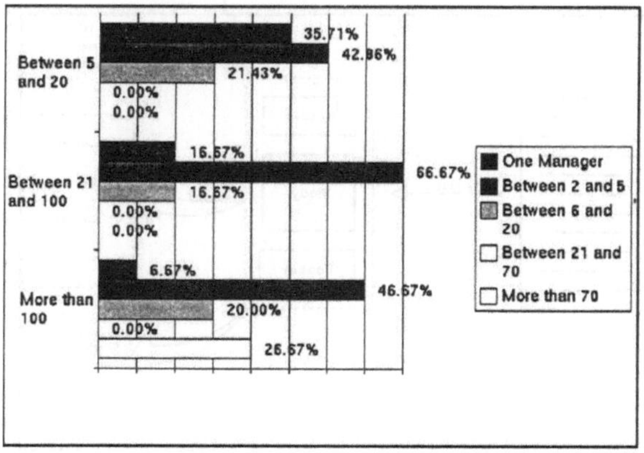

Figure 8: Number of Managed Computers X Network Management Team Size

4. THE NETDOM SYSTEM

4.1 Document Standardization

The first problem to be considered concerning network documentation is how *to build standards documents.* Note that document standardization should be determined inside a network management domain, i.e. a set of documents describing similar elements (devices, cables, services etc.) in an institution have no reason to have a different structure among themselves. Suppose you have a company named A and other company named B. All documents describing FTP servers, for instance, in company A share the same structure. In company B we have the same framework. Nothing imposes that documents describing FTP servers in company A and B also share the same structure, since they describe the same managed element.

To achieve this level of standardization and generalization, a mechanism of validation has to be properly defined.

4.2 Validating Documents

To proper classify a given document as valid or not valid, a system should use a set of formation rules defined as a grammar. Formally a grammar is a set of rules stored in the document file or in separate file(s). Each rule specifies a name and a formation pattern for a fraction of the document. Figure 9 exemplifies a grammar specification scheme.

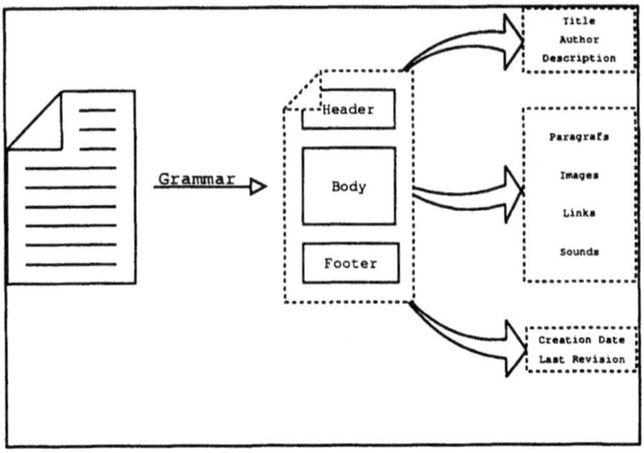

Figure 9: Grammar Specification Scheme

Each document should have its own grammar. Each grammar is defined according to what manageable element it pretends to rule documentation for. This structural organization allows network managers to define precisely what information should be documented and how this information will figure in the final version of the document itself.

Figure 10 outlines one of the possible abstractions for network management. Manageable elements are divided into three distinct classes: *Connectivity Elements*, *Equipments* and *Users*. Analyzing this abstraction as a graph, each leaf, a specific manageable element, should have its own document and, consequently, its own grammar.

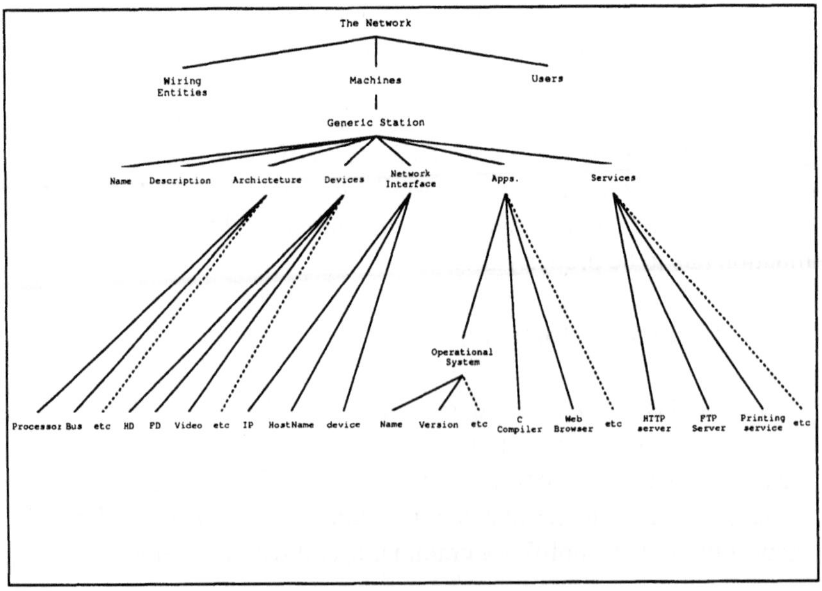

Figure 10: Network Abstraction

4.3 Information Management and Retrieval

Network documentation database contains sensitive information about a company. Providing uncontrolled access to this information base may severely compromise the company's data integrity and availability. Besides, service performance and procedure control may also be affected. Thus, any good network documentation tool must provide a high and incisive level of security.

Another approached problem are the different roles each network management employee plays in a company's hierarchy. A network management coordinator has access to much more information than a connectivity specialist, for instance.

4.3.1 Authentication Engine

Before a user can search a given document and access it, the system must check if this user is a trusted user, i.e. a user that has an explicit permission from the network manager team chief to access network documentation database.

Although this individual mechanism of authentication addresses the problem of who can access the system, a simple yet useful way to group system users is still desired.

To solve this problem, a by-category mechanism should also be defined. Each category should have its own basic permissions regarding general documents. The network management team chief or some other employee leading the network management team should properly set up all this attributes. This user is pre-defined as being the super-user of the proposed system.

It is important to emphasize that a single user may be part of more than one category at the same time.

4.3.2 Document Views

Although by-user and by-category mechanism define the way users and system interacts, users-documents interactions have not been defined yet.

A viewing mechanism may be seen as an information filter. Different users need different information from each document. There are no reasons to allow full view of a document if a user needs just one little part of it.

To provide this level of control, a viewing mechanism bases its decision in three parameters, in the following order:

- User's category;

- Document default permissions to this category;
- Special permissions given to the user by the document's author and/or network management team chief or other defined super-user.

Three permissions are enough to satisfy the above-mentioned requirements. *Read* permission allows a user or all users of a category to retrieve a given section. *Write* permission allows a user or all user of a category to change document contents. *Special* permission allows a user to grant permissions he owns to other users, except special permission.

4.4 System Description

Figure 11 gives a general overview of the system's architecture.

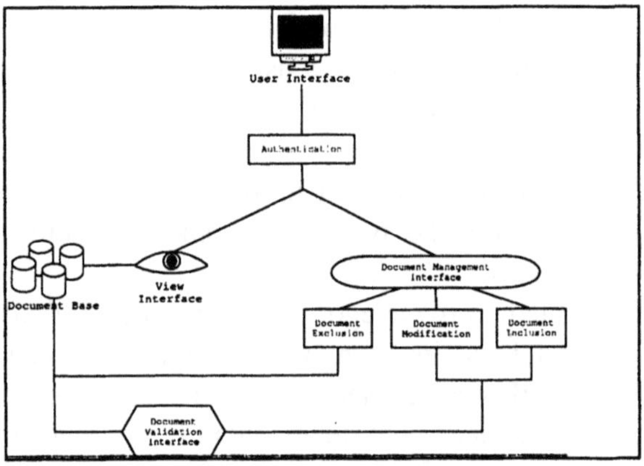

Figure 11: NetDoM Architecture

Authentication interface is the top-most system structure. This prevents unauthorized users to access network documentation database and to compromise network and company sensitive data. Once authenticated, a user can make four different operations:

- *Document Search and Retrieval:* this operation involves *view interface*. According to the user category in the system and to a given document, search results will be displayed. When the user retrieves this document, view interface applies information filtering and the final result will be a content-filtered document according to the user.
- *Document Inclusion:* To properly include a document in network documentation database the system must check it against its

grammar type. This is done by *document validation interface*. Document inconsistency is treated, therefore, in inclusion. This avoids *document base* overload.

- *Document Modification:* As in document inclusion, document modification operation may alter a valid document in such a way that modified document becomes invalid. This problem is addressed by present modified document-to-document *validation interface*. If the document keeps its valid state, old version will be removed from document *base* and replaced by new document version.
- *Document Exclusion:* Exclusions requests are processed directly by document *base*. Security implications of this operation do not represent a drawback, since authentication *interface* has already identified the user to the whole system. So, if a user does not have write access to a given section, this user can delete no documents of this section.

5. MODELING

An object-oriented approach has been used to model the system. Class diagram is presented in figure 12.

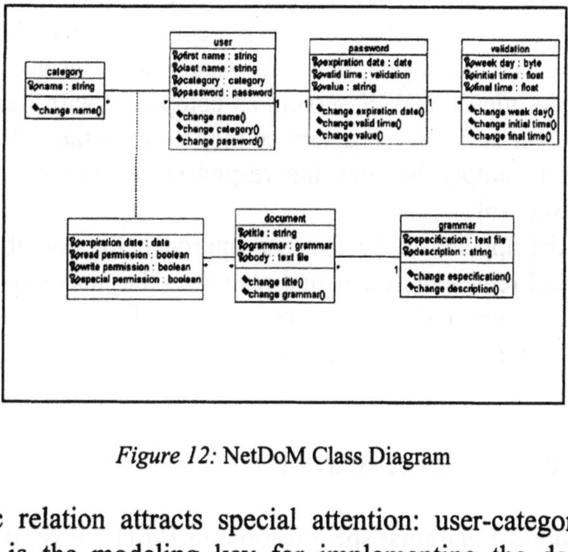

Figure 12: NetDoM Class Diagram

A specific relation attracts special attention: user-category document. This relation is the modeling key for implementing the document view mechanism and all security-related features concerning document accessibility. Note that association class used has an attribute named expiration *date*, which determines another security extension parameter to the network management team's chief and/or the document authors.

A password-validation class relation achieves another possible extension to the authentication engine. Each validation *object* determines a valid range in which a given password is allowed to access the system. In such a way it is possible to restrict to each user a valid access time to the system.

A complete reference on system modeling issues may be found in **[1]**.

6. DOCUMENTING HTTP SERVERS

The main objective of this case study is to properly develop a grammar for a HTTP server. Much of this case study is based in **[3]**.

6.1 Information Domain

Since the main interest is to generate a document for later consultation, it is not interesting to store inherently dynamic information in generated document.

The MIB definition for HTTP server is discussed in **[10]**, **[2]** and **[6]**. Although in section 2.2 we have mentioned that SNMP-based tools are not well suited to document networks, MIBs are a good source for grammar definitions.

Kalbfleisch **[10]** affirms that a WWW server can be viewed from several perspectives when assigning management responsibilities. These perspectives are named the operational *model* and service *model*. The operational model views WWW servers as computers with hardware, disk, OS, and web server software. The service model views the WWW server as a black box that simply handles the responses to requests from clients located on the web. **[6]**

Service model appears to be the best approach for our objective since operational model represents the actual resources that make up the machine so that it can be monitored from the perspective of resource utilization **[6]** and this kind of information is inherently dynamic.

6.2 Grammar Specification

Figure 13 shows a grammar specification for HTTP Server documents.

Grammar: HTTP Server
Document: Header, Body;
Header: *Author, Organization;
Author: User;
Organization: String;
Body: Server, Version, Installation, Deactivation, DocumentRoot, *ConfigFile, Port, #Module, ServerName, #Administrator, UserDirectory, *IndexFile, AccessMode,
Notes;
Server: *String*;
Version: *String*;
Instalation: *Date*;
Deactivation: *Date*;
DocumentRoot: PathName;
ConfigFile: PathName, FileName;
Port: *String*;
Module: *String*;
ServerName: *String*;
Administrator: *User*;
UserDirectory: PathName;
IndexFile: FileName;
AccessMode: *String*;
Notes: Text;
PathName: *String*;
FileName: *String*;
Text: * *String*;

Figure 13: HTTP Server Document Grammar

Entries proceeded by ``*" figure one or more times in a given token. Thus, text token contains one or more strings, for example. Entries preceded by ``#" figure zero or more times in a give token. Thus, body token contains zero or more module entries. Emphasized entries in figure 13 represent core data types.

7. CONCLUSION

Even tough a considerable effort has been made in the study of computer network management issues, network documentation has been neglected. To solve this problem, network managers begun to make use of alternative techniques. This fact may be verified by the results and analysis of an on-line survey developed and described in section 3.

As expected, such techniques presented several drawbacks, including sensitive information endangerment, accessibility and retrieval problems and lack of information restriction methods and document standardization. We have proposed an information system to address all the above-mentioned problems.

Possible extensions to this work include the study of security-related subjects about the described information system, analysis of storage techniques for semi-structured documents, use of mark-up languages such as XML and SGML to development of document-based information system.

REFERENCES

[1] Martin Fowler and Kendall Scott, UML Distilled: Applying the Standard Object Modeling Language, Addison-Wesley, 1997,

[2] Harrie Hazewinkel and Eric van Hamgstum and Aiko Pras, Definitions of Managed Objects for HTTP, Available on line at http://wwwsnmp.cs.utwente.nl/nm/research/results/publications/pras/Internet-drafts/mib-http.txt, 1996

[3] Aiko Pras and Harrie Hazewinkel and Eric van Hengstum, Management of the World-Wide Web, Proceeding of 15th. Brazilian Network Symposium, SBRC,1997,

[4] M. Fedor and M. Schoffstall and J. Davin, RFC 1098 -- A Simple Network Management Protocol (SNMP), 1990

[5] K. McCloghrie, RFC 1213 -- Management Information Base for Network Management of TCP/IP-based Internet: MIB-II, 1991

[6] C. Kalbfleisch, RFC 2039 - Applicability of Standards Track MIBs to Management of World Wide Web Web Server, 1996

[7] V. Konstantinou and P. Morse, Electronic Documentation System: Using Automated Hypertext Techniques for Technical Support Services, Proceedings of the 10th Annual International Conference on Systems documentation, 1992

[8] M. Rose and K. McCloghrie, RFC 1155 -- Structure and Identification of Management Information for TCP/IP-based Internets, 1990

[9] K. McCloghrie and M. Rose, RFC 1156 -- Management Information Base for Network Management of TCP/IP-based Internets, 1990

[10] H. Hazewinkel and C. Kalbfleisch and J. Schoenwaelder, RFC 2594 -- Definitions of Managed Objects for WWW Services, 1999

[11] Highline Community College, System Services: Local Area Documentation, Available on line at http://hcc-web265.highline.ctc.edu/sysserv/newunetg.htm

[12] Centro Nacional de Processamento de Alto Desempenho no Nordeste -- CENAPAD-NE, Configuração da Rede, Available on line at http://www.cenapadne.br/rede.htm

[13] Rede Nacional de Pesquisa -- RNP, RNP Documentos, Available on line at http://www.rnp.br/documentos/

[14] Marcelo V. Vasconcelos, 'Atila, Uma Arquitetura Inteligente e Distribuída para o Gerenciamento Pró-ativo de Redes ATM, M.Sc. Dissertation, 1996.

MIX
Papier aus verantwortungsvollen Quellen
Paper from responsible sources
FSC® C105338

If you have any concerns about our products,
you can contact us on
ProductSafety@springernature.com

In case Publisher is established outside the EU,
the EU authorized representative is:
Springer Nature Customer Service Center GmbH
Europaplatz 3, 69115 Heidelberg, Germany

Printed by Libri Plureos GmbH
in Hamburg, Germany